by
DicQie Fuller, Ph.D., D.Sc.

The Healing Power of Enzymes

60 Fifth Ave • New York, NY • 10011

NEW YORK • CHICAGO • WASHINGTON D.C. • LOS ANGELOS • TORONTO

Reprinted 1999

CIP Data is available
Printed in Canada
10 9 8 7 6 5 4 3 2

ISBN 0-8281-1289-4

Contents

Part I

Supercharge Your Health with Enzymes

Chapter 1

The Miracle of Enzymes

We humans are the great miracles of creation! Like snowflakes, each of us possesses a one-of-a-kind pattern for our own unique state of health. This pattern is known as enzyme potential, it is contingent upon individual makeup, and heredity. For instance, members of the same family may be of various shapes, with divergent health problems and dietary cravings, all because of their personal DNA structure. Sufficent genetic and biochemical data was correlated by the mid 1940s to establish our enzyme potential is inherited (*Genes and Genomes*, Singer and Berg).

Just what are enzymes, anyway?

It is ironic that almost everyone knows the value of vitamins and minerals for good health. Knowledge of the importance of herbs is becoming more widespread as well. Although enzyme biochemistry has been the subject of intensive study, the role of enzymes in supplementation and therapy is not widespread knowledge. Enzymes are proteins taken from food and made by the body. They must be present before any chemical reaction can take place in our bodies. Even vitamins, minerals, and hormones cannot do their jobs without enzymes. Too small to be seen through the most powerful microscope, they are the catalysts, the dynamic power that gives us our ability to function at the highest level of good health. Our bodies cannot exist without enzymes.

Enzymes are protein taken from food and made by bodies.

This book examines those enzymes called **digestive plant enzymes**. For proper digestion to take place, four major and three minor enzymes are required. The enzyme called **amylase** aids the digestion of *carbohydrates* and *starches* (fruit, vegetables, pasta, and bread). Poorly digested carbohydrates/starches ferment in the system, creating gas and other symptoms of discomfort. **Protease** enzymes digest *proteins* (red meats, fowl, fish, and nuts). If improperly digested, protein putrefies in the system, causing indigestion and toxicity. **Lipase** breaks down *fats* and assists in balancing fatty acids. Fats that are not thoroughly digested turn rancid in the system and create pungent odors and poor cholesterol balance. **Cellulase** breaks down cellulose (the fiber found in vegetables and other plant materials). However, the body does not produce cellulase. Hence the synthetic commercial fiber we ingest for the purpose of scouring the colon of unassimilated matter often fosters more intestinal problems. Undigested fiber leaves a waxlike residue in the small intestine and adds to absorption maladies. The most widely known of all enzymes, **lactase**, breaks down the *milk sugar lactose*. If the body is not producing lactase, the lactose in milk is not broken down, resulting in digestive problems. Additionally, some dairy products can cause allergies. Two other lesser known enzymes that break down *food sugars* are **sucrase** and **maltase**.

Enzymes must be present before any chemical reaction can take place in our bodies. Even vitamins, minerals, and hormones cannot do their jobs without enzymes.

What are some common sources of enzymes?

Metabolic enzymes are those produced by the body. They are necessary for all stages of the digestive process. Age and high stress reduce our ability to manufacture these enzymes. **Food enzymes** are preserved intact in raw foods, which is why raw foods are so nourishing. Uncooked fruit and vegetables contain their own enzymes. These foods, so wholesome when raw, completely lose their ability to aid digestion when cooked. A ripening banana is a good example of how raw-food enzymes work. The banana transforms itself from a hard, starchy plant into a very soft, brown and sweet one *through a process requiring the energy from enzymes.* The third source for these miraculous substances is **supplemental digestive enzymes**. Derived from plant foods in a laboratory, they are known as **plant enzymes**. Enzymes cannot be made like synthetic vitamins and minerals. They must be grown in plant form and extracted in a laboratory process. Supplemental plant enzymes, usually in capsule form, are swallowed with food to assist in the digestion of that particular meal. They work throughout the entire digestive system, from esophagus to rectum. *Other enzymes available for human consumption are **animal enzymes**. One of them, **pancreatin**, comes from the pancreas of a slaughterhouse hog or ox. Pancreatin requires an alkaline pH setting to work. It begins working in the latter stages of the digestive process in the alkaline area of the small intestine where fat is digested. Three other enzymes are useful under certain circumstances. **Bromelain**, an enzyme derived from the pineapple plant, and **papain**, an enzyme derived from the papaya plant, are used as meat tenderizers and ingredients in beer. **Pepsin** is an enzyme involved in the breakdown of proteins in the stomach. usually prepared from the stomach of pigs and requires an acid pH setting.

Because of their ability to aid digestion throughout the entire digestive process as well as support the endocrine system, plant enzymes are generally more useful to the body than any other type of enzyme supplement.

Why do we need enzymes?

We need enzymes when we eat food that has been cooked, frozen, processed, or altered in any way. Enzymes are destroyed by the heat produced when we boil, microwave, or pressure-cook food. If the temperature rises above 118 degrees Fahrenheit while steaming, even that process can destroy enzymes. Certain methods of juicing, which produce heat through friction, can suppress the enzyme action when we digest our food.

Eighty percent of our body's energy is expended by the digestive process. If you are run down, under stress, living in a very hot or very cold climate, pregnant, or are a frequent air traveler, enormous quantities of extra enzymes are required by your body. Because our entire system functions through enzymatic action, we must supplement our enzymes. Aging deprives us of our ability to produce necessary enzymes. The medical profession tells us that all disease is due to a lack or imbalance of enzymes. Our very lives are dependent upon them!

Energize your life with Nature's lifesavers.

These delicate substances are present in all living cells, whether vegetable or animal. Technically, enzymes are energized protein molecules. Our bodies can obtain them in only two ways — by producing them (endogenous enzymes) or by acquiring them through the food we ingest (exogenous enzymes). There are two problems that occur with each source: As you get older, your body becomes much less efficient and, therefore, produces significantly fewer enzymes than it did as a youngster. Second, the greatest part of the food you eat is cooked, which kills all enzymes before they ever reach your mouth. The results are that unless you are a child or you eat only raw foods, you do not get enough enzymes to replenish your supply to remain healthy and energized. Furthermore, because enzymes are protein molecules, the body needs a constant supply of amino acids from our diet to continue to replace lost metabolic enzymes. If the digestive process is not complete, the supply of amino acids decreases, further hindering the body's ability to replenish its metabolic enzymes.

These wonderful little lifesavers serve your body in two important and distinctly different ways. First, they break down the nutrients contained in food into chemical substances fine enough to pass through the lining of the digestive-tract cells. From there, they enter the blood stream. The entire process is fascinating, and I plan to cover it in more detail in Chapter Two. Your body considers digestion its most important duty. When a newly eaten meal enters the stomach, digestive enzymes are pulled from all systems of the body to get to work immediately on digestion. However, this same enzyme energy is also needed elsewhere to repair, regulate, and reactivate the other systems of the body. These systems have no choice but partial shutdown during digestion. One remedy is to eat only raw foods complete with their own supply of digestive enzymes. However, there are only enough enzymes contained in the raw food to digest that food. The other alternative is to supplement your diet with enzyme capsules derived from plants. This will be covered extensively in the next chapter.

It is important to remember that poorly digested carbohydrates—fruits, vegetables and starches—will ferment in the intestinal tract. Fats—dairy products, oils and fried foods—turn rancid. Animal proteins—like fowl, fish, and nuts—putrefy in your system if undigested. Is it any wonder that we are an entire society suffering from constipation, gas, indigestion, and halitosis?

The other major role of enzymes is to keep the metabolism working at full throttle. This includes burning fats, feeding cells, and releasing energy wherever needed. Every bodily function, from building new flesh, muscle, bone, glands, and nerves to those which rid the system of toxins, depends completely on enzymatic activity. Skin glowing with health, an efficiently operating colon, liver, heart, and brain, fully functioning lungs, kidneys and hormones all rely on enzymes. It can truly be said that enzymes are Nature's lifesavers!

Dr. Peter Rothchild, M.D., a world-renowned researcher and Nobel Prize nominee in physics writes, "Evidence of the effectiveness of enzymes taken orally is beginning to overcome skeptics. Many of these studies show enzymes, when taken orally, demonstrate benefits against circulating immune complexes, rheumatic disorders and acute immune diseases."

Other highly respected scientists, physicians and researchers who believe that chronic disease, cancer and ill health are linked to enzyme deficiency are: Dr. Franz Berge of London; Dr. R. A. Holman of the University of Wales; Dr.

Van Fotter, Professor of Cancer Research, University of Wisconsin; Dr. Harold Manner, Loyola University; and Professor William T. Salter, Yale Medical School. Nobel Prize recipient Dr. James B. Summer states, ". . .the getting old feeling after forty is due to reduced enzyme levels throughout the body. Young cells contain hundreds more enzymes than old cells. Old cells are filled with metabolic waste and toxins."

D. A. Lopez, M.D.; R. M. Williams, M.D. Ph.D.; and M. Miehlke, M.D., write in their book, *Enzymes: The Fountain of Life,* "We feel that the therapeutic use of enzymes is likely to become a major form of medical treatment in the future. Their therapeutic potential is enormous! We believe that enzymes are the treatment of the future!"

Comparing your enzyme inventory to a bank account.

Dr. Edward Howell, a brilliant researcher who began studying food enzymes and human health in 1930, wrote a wonderful book I highly recommend called *Enzyme Nutrition,* Avery Publishing Group Inc., 1985. In it he suggests that the body can be considered an "enzyme bank account." Trillions of cells work constantly to maintain an acceptable balance of enzymes to keep the body working. The only other backup source we have comes from the food we eat. What happens, Dr. Howell asks, if we experience a big enzyme drain such as contracting a virus, some strenuous exercise, an emotional crisis, breathing unclean air, staging a temper tantrum or eating a fat-laden meal? His answer: "The balance in your enzyme bank account drops so much that your body faces enzyme bankruptcy." The solution: Make a quick save-the-day "deposit" to keep your systems in operation. Its source: Plant enzymes taken in capsule form to rescue your digestive process, fortify your tissues, and fuel the metabolic enzyme activity that actually keeps you alive and functional. Unfortunately, eating some raw fruits and vegetables will not provide an adequate "deposit" as they contain only enough enzymes to digest their own particles. No surplus exists.

Time out for a minute! Why haven't I heard about enzymes?

If you are like most of my clients when they first hear me speaking enthusiastically about the endless roles played by enzymes, you are beginning to wonder why you have not seen or heard much press coverage about them. The answer is probably because there are not yet enough knowledgeable people around to supply the necessary information. In the case of vitamins and minerals, you do not have to look far to find experts who can explain the way these important substances support our health. Many opportunities exist for reporters who want to write educational articles about the value of diets that are rich in vitamins and minerals.

But how many enzyme specialists do you know? Furthermore, would the prospect of reading a newspaper article or watching a TV documentary on enzymes fill you with excitement? Probably not. Yet the healing power of enzymes has been documented for decades. We would all benefit knowing how, when correctly used, they can do wonders for our minds and bodies. We have briefly touched on the importance of enzymes for good digestion, and Chapter Two will cover that in more detail. For now, let's take a brief look at other functions of the body and their dependence on enzymes for efficient operation of the many metabolic processes. Each chapter that follows will give you greater details on each topic.

You are probably reading this book for one of two reasons. You care enough about your own health or someone who cares about you gave it to you. No matter which way you came to read about enzymes, you are entitled to a clear explanation of how they work and why they are so important. Many enzymes are on the market today, each with its own purpose.

First, we will look at *industrial grade enzymes*, such as those used in cleaning agents. These enzymes act as grease-eaters. In this category, there are other single-purpose enzymes designed to do one specific thing, such as break down a particular type of grease. There are also *commercial grade enzymes*. These are sold in large quantities and are used for baking, tanning leather, or fermenting the malt used to brew beer.

The focus of this book is *nutritional pharmaceutical grade plant enzymes*. These have a broader spectrum of use and come in a more purified and stable form. They are the ones I have studied most extensively. I personally use these enzymes, and only these. Here are my reasons for doing so. I wish I could say that all packaged, nutritional enzymes are completely safe and free of contaminants. Sad to say, that is not the case. Some companies do not use the purest products, so they can sell at lower prices. The enzymes I use are those grown for a specific function—efficient digestion. I am not interested in those commercial grade enzymes found in the bakeries and breweries. I recommend that anyone genuinely concerned with improving his or her health should avoid them as well. Many of you have not had access to a high-quality source. I have met longtime users of enzymes who have had adverse reactions to them. This is because these products are impure. Some kind of filler has been used to allow the manufacturers to sell at lower prices. Only those enzymes with no fillers should be ingested to insure your maximum health and well-being. If it is necessary for you to take those with fillers, make sure the fillers come from a product that contains nutrients. It is a well-known fact that not everything is "natural" even if labeled that way. Synthetic vitamins can be legally labeled "natural" if they contain a small amount of natural food ingredients. People often think they are taking high dosages of absolutely pure vitamins or enzymes from an "all-natural" source. The truth is that if these high-strength capsules were derived entirely from a perfectly pure, natural food source, one of them would be the size of a table. That would be difficult to swallow! This is why I want the public to be aware of what they are getting for their investment.

Plant enzyme supplements can be purchased in health food stores, through multi-level marketing groups or mail-order, and from health care professionals. They will vary in strength, ingredients and the amount of fillers they contain. There are even claims that certain vitamins or foods are enzymes. Do not confuse plant enzymes with those made from animal products. Of all enzymes used for increasing our level of good health there is no such thing as a bad enzyme supplement. Remember, each of them works differently and may be intended for something other than digestion. You can find more information on this in Part Five of this book, the Question and Answer section.

Careful . . . Watch it! Oh, never mind . . . 10,000,000 of your cells just died!

To illustrate the importance of enzymes, did you know that in less time than it takes you to read this sentence, and well before you finish the paragraph, ten million of your cells have died? Are you aware that there are at least one hundred trillion cells at work in your body right now? Scientists have identified at least twenty-eight hundred different enzymes. They estimate that there are more than three thousand active enzymes at work in our bodies. It takes thirteen hundred enzymes to make one cell!

Enzymes are pure protein. They are catalysts in the combination of different molecules. Without them, a single cell consisting of more than one hundred thousand different chemicals could not be. Your DNA molecules receive their ability to do all the tasks on the cellular level because of the presence of enzymes. As I stated earlier, no less than twenty-eight hundred enzymes are necessary for the life process to take place. The required nutrients our bodies use to make metabolic enzymes come from the food we eat. That is, if we digest our food thoroughly and use the nutrients efficiently.

Forty-five nutrients (in their proper amounts) derived from the food we eat nourish each and every cell. These nutrients are:

19 minerals, in prescribed amounts
13 vitamins
9 amino acids
1 protein
1 fat
1 water
1 carbohydrate
Plus 1300 enzymes

Way to go . . . You just made billions! (Of red blood cells, that is.)

When I suggest that there is an exact amount of vitamins and minerals needed to maximize good health, I mean that mega-dosing can create more problems than consuming too few nutrients in some instances. What regulates all this is the delivery of enzymes. For example, there are billions of red blood cells made from bone marrow each second. In order to build these red blood cells, a form of iron called heme-iron is necessary. We need to eat protein foods in order to obtain heme-iron. However, if the protein food is not properly digested, our body will not extract the nutrients required. Inefficiencies in our digestive system typically allow only about 15 percent of the heme-iron present in meats, and only about 3 percent of the heme-iron present in plant foods, to be absorbed by our bodies. Improving the efficiency of the digestive system is vital to obtaining nutrients such as heme-iron. Digestive enzymes are what allows the body to properly digest and assimilate the nutrients we consume.

The importance of proper digestion is mind boggling. Every function must be perfectly synchronized with every other function. When we lack a particular enzyme, vitamin, or mineral the resulting imbalance causes disease.

If your intake of calcium is too low, you will experience hair loss, brittle finger and toe nails and show all the symptoms of those who have bone problems. Too much calcium can form bone spurs and cause blood disorders. What happens when you do not have enough enzymatic action in your body? Once again, it takes most of our energy just to digest our food. Digestive enzymes are the delivery mechanism in the body. Not only do they assist in the breakdown of our food, they deliver the nutrients to every part of the body.

Absorption cells called *villi* are present in the small intestine to aid with digestion. If all the villi in our body were laid out end to end, they would measure about half the length of a football field. Villi replace their cells every two days. With aging, we lose our ability to replace these cells, so they can no longer absorb nutrients for delivery to other cells. Where an imbalance occurs, such as Celiac disease, the villi fold over and are not able to absorb the nutrients. The next step is that undigested food, such as protein, begins to clog the system. Remember, undigested protein begins to putrefy and will eventually cause health problems.

The importance of proper digestion is mind boggling. Every function must be perfectly synchronized with every other function. When we lack a particular enzyme, vitamin, or mineral, the resulting imbalance causes disease. We are now considering the following facts due to the unhealthy eating habits that dominate our society: If you have digestive problems, you will have trouble rebuilding cells, which leads to an array of health disorders. My own experience in working with medical doctors and researchers has proved this to be true. You may think it is too simplistic to conclude that illness is caused by inadequate digestion, but I am convinced this is true.

Is there a way to know what enzymes are deficient in my body?

The obvious symptoms of poor digestion are burps and belches, excessive gas, heartburn, nausea, constipation, feeling tired after eating or an allergic reaction to a particular food. Another way to discover a deficiency is through observing which foods you crave or eat the most. These include caffeine, chocolate, refined sugars, and cooked or processed foods.

I refer you to Part Three: "The Enzyme Effect on Body Shape and Weight" for further information on enzyme deficiencies. Now may be a good time to determine your body type.

There are four basic body types, each associated with a specific enzyme deficiency. A brief description of the four types follows. If you gain weight uniformly, meaning you are not carrying any more weight on one part of your body than another, and your buttocks are noticeably rounded, you desire sweet foods. Among these items are cookies, cake, potatoes, pasta, fruit, coffee, and chocolate. This is known as the Para Body Type (Type One) and indicates a lack of the amylase enzymes.

If most of your weight is carried in your buttocks and thighs (your shoulders are narrower than your hips) and you like strong-flavored foods, you are an Estro Body Type (Type Two). You prefer ethnic foods; spicy, salty and rich, and you are lacking in lipase.

If you gain weight or carry your weight in the upper torso, (stomach, shoulders, back and chest) and your buttocks are flat, you are a Supra Body Type (Type Three). For you, no meal is complete without some form of protein such as steak, fish, fowl, pork, or eggs. This denotes a deficiency in the protease enzymes.

Some of us have the same body as adults that we had in junior high and high school. This body type can be boyish and wiry or soft like a baby, shaped without defined curves. You eat more dairy products than anything else, unless you have already developed lactose intolerance. This is the Neuro Body Type (Type Four); deficient in lipase, lactase, and amylase.

If the food you eat does not contain the right balance for your body type (genetic need):

(A) it will not give you the correct nutrients for your body type;
(B) it will not be burned as fuel, but will store as fat;
(C) it will create toxicity for your cells and will lead to fatigue.

It is possible to be a combination of more than one body type. Our heredity dictates our individual body's strengths, weaknesses and shape. We also inherit our "enzyme potential," an unbalanced system and need for specific enzymes. This will be covered more extensively in Part Three of this book.

Chapter 2

New Life for Your Digestive System

What is wrong with my digestion? Why can't I eat the foods I used to eat? These questions have been asked of health-care practitioners throughout the last few centuries. Most people never receive an answer or a solution to their problems. To say that the body can easily digest and assimilate cooked food, as so many nutritionists do nowadays, may one day prove to be completely wrong. Cooked food passes through the digestive tract more slowly than raw food. Therefore, it tends to ferment and putrefy, throwing toxins into the system. These cause gas, heartburn, headaches, eye problems, allergies, and a host of other serious conditions.

Supplemental plant digestive enzymes are used in the gastrointestinal tract in order to break down food. Under optimum conditions, this process accomplishes three things: (1) Food is completely digested; (2) nutrients are absorbed into the blood; and (3) waste products pass through and are eliminated. Plant-based enzymes are much more beneficial than animal enzymes because they activate more quickly. They begin their work immediately after ingesting, while food is still in the upper part of the stomach. Sometimes, they activate while dissolving in the esophagus.

We are very aware of papaya and pineapple as sources of enzymes that break down protein. They require a pH that is not commonly needed by the body. Rather, it is aspergillus plants (single cell plants) which come from natural foods that are the most worthwhile source of plant enzymes. They perform most efficiently in humans, insuring the digestion of fats and carbohydrates as well as proteins. Aspergillus enzymes serve a dual purpose: They prevent waste from entry and storage in our cells, and cleanse cells of old waste matter.

Your digestion is only as efficient as your stomach's ability to begin the process. To be healthy, your system has to assimilate food, distribute the nutrients, and eliminate waste through the bowels. All this must happen without discomfort or gastrointestinal stress. Earlier, we learned that **carbohydrates can ferment, fats turn rancid, and proteins putrefy**. This undigested material is known as food remnants.

A major problem with undigested food is that it often passes into various parts of the body where it is deposited as waste. This problem manifests itself as high cholesterol, calcium deposits, arthritis, cellulite, and more seriously, as plaque in the arteries.

Let's review once again what it is that digestive plant enzymes do.

Proteases: breaks down protein
Lipases: breaks down fats
Amylases: breaks down carbohydrates/starch
Cellulases: breaks down fiber
Lactases: breaks down milk sugar
Maltases: breaks down malt sugar
Sucrases: breaks down sucrose (refined sugar)

It needs to be understood that our bodies do not manufacture cellulase. We must ingest it. In addition to digesting fiber, cellulase transforms into cellulose and glucose and short-chain fatty acids.

With food allergies becoming a common problem and more genetically altered foods appearing on the market, we must make enzymes a part of every meal.

Gallbladder

Burping, belching, a bloated feeling, hiccups, intestinal gas, regurgitation, discomfort in the rib cage area, nausea, diarrhea, breathing difficulty, overall exhaustion, heartburn, anxiety, rapid heartbeat, headaches, and constipation are the results of a stressed gallbladder. Lumped together, we describe them as "indigestion." Gallbladder problems are triggered by the inability to break down fats or excessive alkaline or acid pH in the digestive system. If you have any of the following symptoms you may have gallbladder stress or disorder.

Nausea or diarrhea after eating, immediately or hours later
Headache before or after eating
Nausea relieved by eating or a bowel movement
Burps and belches
Fatigue after eating
Indigestion
Diarrhea followed closely by constipation
Pain or tenderness in the right upper abdominal area
Bloating or tenderness at the base of the sternum (breastbone)

The gallbladder becomes stressed due to poor eating habits. When too many alkaline foods are ingested, more bile is produced. This starts peristalsis, hence the probable occurrence of nausea or diarrhea. Alkaline foods include most fruits, vegetables, some grains, fruit juices, Brazil nuts and all seeds. Fatty and highly seasoned foods such as ice cream and most other desserts, Mexican, Italian and Chinese cuisines, and chocolate may also irritate the gallbladder.

Gallbladder in Relation to Individual Body Type

In Body Type One, gallbladder stress occurs through cravings for alkaline-rich carbohydrates. Body Type Two suffers the same thing but for a different reason—poor fat digestion. If a low-protein, high-carbohydrate diet is followed, insufficient bile is produced. If this happens, the gallbladder cannot empty itself and the liver is not stimulated to produce the correct amount of bile. Fats remain in such large particles that metabolic enzymes cannot combine readily with them. Hence, fat digestion and absorption are incomplete. The undigested fat combines with any iron or calcium contained in the food. These form insoluble soaps, which prevent minerals from reaching the blood. These hard soaps can cause constipation, anemia and bone loss. In addition, the lack of bile acids prevents the absorption of carotene, Vitamins A, D, and K, plus essential fatty acids. This all leads to a multitude of deficient disorders. By now you are beginning to understand that eating healthy food is not enough. If it is not being digested properly, or it is inappropriate for your body type, many problems occur. Continued depression of the immune system is one of the most serious. This is where multi-enzyme supplements can be very effective.

Hiatal Hernia

Hiatal hernia is a weakened, stretched or protruded muscle forcing the stomach through a hole in the diaphragm. This interferes with the production of hydrochloric acid because the vagus nerve is pinched. Food cannot enter the stomach. This causes discomfort and pain. Undigested food particles become exposed to bile. Bile, an alkaline substance, acts as a strong soap. The delicate balance of chemicals is disrupted, and the bile washes away the protective coating of mucous in the stomach which guards against extremes of acidity and alkalinity. Cells work overtime to produce extra hydrochloric acid to counteract the excess alkaline bile. The area now unprotected by the mucous coating becomes an irritated lesion, or ulcer. If this happens to the stomach, where good digestion must begin, the entire digestive process is undermined. Difficulties you might encounter with hiatal hernia include belching, bloating, intestinal gas, regurgitation (a burning sensation which feels as if the food has not gone all the way down), discomfort underneath the rib cage, nausea, diarrhea, constipation, difficulty with breathing, exhaustion, pressure and pain in the left side of the chest, anxiety, rapid heartbeat, and headache! Often, patients in emergency rooms think they are having a heart attack, when it is gastric distress or hiatal hernia. There is a good deal of similarity between gallbladder and hiatal hernia symptoms because they are closely connected. A hiatal problem, first triggered by poor fat digestion, puts undue stress on the gallbladder. Likewise, gallbladder problems can be the result of a hiatal hernia, depending upon what types of foods are ingested, and whether disruption of the acid/alkaline balance takes place. Later in this chapter, I will cover the importance of good acid and alkaline balance.

What is wrong with my digestion? The best answer to this cry for help is plant enzymes. Research has shown that a combination of the proper enzymes restores pH balance throughout the digestive tract. In the case of the hiatal hernia, if supplements are taken with the meal the enzymatic action could begin even in the protrusion. Introduced enzymes would stabilize the balance between alkalines and acids.

Food Allergies

Food allergies are nothing more than the body's toxic reaction to substances it cannot break down. The word **allergy** means an abnormal and individual hypersensitivity to substances that are ordinarily harmless. Numerous case histories show positive results from the use of digestive enzymes as a treatment for food allergies. Common sense shows when we supplement with enzymes, we can boost a digestive process that no longer produces enough enzymes on its own. When we do this, our food allergies disappear. Because we all have different body types, we will differ in the amount of time and the number and variety of enzymes we will need to be able to safely reintroduce irritant foods into our diets.

As an example, I will discuss the importance of good digestion, which begins with the pancreas. The pancreas produces enzymes for digestive action. If the pancreas has undergone too much stress from overuse through improper eating, it becomes overwhelmed. It can no longer meet the excessive demands placed upon it. When the pancreas loses its ability to produce pancreatic proteolytic enzymes, this creates an insufficient quantity of various enzymes. One consequence is poor digestion of proteins into amino acids. Proteolytic enzymes are made up of amino acids, and if amino acids are deficient, these enzymes which prevent and eradicate inflammation will also be ineffective. This sets off a chain reaction that reduces enzyme production elsewhere in the body. Other areas adversely affected by these deficiencies are hormonal activity; excessive demands for vitamins, mineral, and trace minerals; lowering of immune defenses; infection; and a general feeling of malaise. The end result is unusable protein molecules that are intended to fight inflammation. Instead, they are absorbed through the intestinal mucosa and circulated in the blood, entering tissue in partially digested form.

These partially digested protein molecules are known as peptides. The body regards them as invaders. They cause a kinin-mediated inflammation. Kinin chemicals cause pain when they come in contact with nerve endings or tissues of specific organs.

Nature has charged the pancreas with the essential task of providing enzymes to control inflammation, regardless of its cause. It could be due to a cut, a bruise, or a chemical reaction in the food we ingest. Conventional medicine has produced chemically based anti-inflammatory drugs that reduce inflammation. However, these drugs do not cure it.

If we are unable to break down our food into small enough particles, we will suffer from inflammation of the organs as well as the joints.

It matters greatly to our overall health what foods we eat, or whether we smoke or drink, because this has much to do with our own nutritional state. If we are unable to break down our food into small enough particles, we will suffer from inflammation of the organs as well as the joints. This can manifest as edema or tissue swelling, histamine, sinus and allergy problems, and a general loss of good health. We become dependent on antihistamines, nasal sprays, chemical tranquilizers, antidepressants and aspirin.

Anyone who is nutritionally deficient, infected, toxic, or who has food addictions will suffer the consequences. These develop into chronic degenerative illnesses unless something is done to stop the progression. I firmly believe that enzymes are necessary for the good health of everyone. My clients with chronic illnesses have all been afflicted with poor digestion for a significant number of years. Prevention is the best prescription. Remember, it means to stop something before it happens!

Gastritis Relief

The three most important enzymes for treating gastrointestinal disorders are large amounts of amylase and lipase, and a small quantity of cellulase. Some herbs are helpful as well. I personally recommend marshmallow root. This herb works beautifully in assisting the body to manufacture the materials that build new tissue. It has the adhesive property that allows it to surround and expel foreign or toxic matter. Marshmallow root has a soothing effect on inflamed mucosal membrane tissue in both the gastrointestinal and respiratory tracts. It helps to normalize mucous secretions and remove excess accumulations of uric acid. Another very effective herb is Gotu Kola. Known as a longevity herb, it is rich in vitamins and minerals. Marshmallow root used in conjunction with Gotu Kola works even more efficiently. Any discussion of herbal remedies for all kinds of irritation cannot leave out prickley ash. Prickley ash is a powerful treatment for the bowels, lungs and stomach. Many herbs, when used in correct proportion with the right enzymes, can produce a very calming effect if emotional stress is the cause of your gastritis. One of them is valerian root. Others are hops, skullcap, and wild lettuce. To normalize a crisis, whether it is emotional or physical, you continue to ingest the basic enzymes with each meal.

How to Heal Ulcers

The peptic ulcer occurs most commonly. It is defined as a loss of the tissue lining of the lower esophagus, the stomach, or the duodenum. Less serious, but often confused with ulcers, are lesions. They involve the muscular coat but do not constitute a complete loss of it. These lesions are not as painful as ulcers. While it is known that gastric acid and pepsin are partially responsible for ulcer formation, it is unknown why mucosal resistance to them should become impaired. Many times ulcers are caused by bacteria. There are many theories about how genetic makeup and environmental conditions can cause a peptic ulcer. We do know that they get much worse from eating a diet of hot, spicy foods.

Psychological stress alters gastric functions. We know that stress can create ulcers. A stress ulcer differs clinically from a chronic peptic ulcer. It is more acute, but not as painful. Stress ulcers can be triggered by severe trauma or surgery. Some ulcers are caused by overuse of certain drugs, such as aspirin or alcohol. The symptoms include burning or gnawing pain, cramping or aching. This pain is described as coming in waves, each of which can last several minutes.

In my experience, ulcers can be treated very similarly to other gastric problems. Protease, remember, digests bacteria that does not belong in the system, so it would naturally be beneficial for a bacteria-induced ulcer. As described earlier, the same herbs and enzymes used in combination with one another would also prove to be very effective. These herbs and enzymes can actually help to replace the mucosal lining. I have formulated a product that aids in the replacement of the mucosal lining over time.

This solution includes treating the body with love and care so chronic health problems will not have an opportunity to develop. It is very wise to start your children on enzymes at an early age, rather than waiting until they have contracted an illness of some kind.

Diarrhea

I find it very interesting to hear what the average person thinks of as normal and natural concerning bowel elimination. When clients call me, complaining about diarrhea, I ask them about their symptoms. Is it an explosive, uncomfortable diarrhea? The kind that can be serious due to loss of electrolytes? Are you spending a great deal of time on the stool? Or have your bowels merely changed, are they less thickened or hard? Do you go to the bathroom more often? There is a difference between diarrhea and a loose stool.

If foods are digested correctly and the body is able to get from them the nutrients it needs, you would never have diarrhea.

Diarrhea can be dangerous when the rapid evacuation of fecal matter results in the loss of water and electrolytes and other essential substances, like potassium. This produces an acidosis that creates further problems. Besides frequent bowel movements, there are usually abdominal cramps and an overall weak feeling. Sometimes the stools are coated with blood, or they contain mucus. When we change our diets and eat more raw vegetables and fruits, this creates looser and more frequent stools, a condition not to be confused with diarrhea.

At the onset of diarrhea, take a natural product designed to control it. If it lessens the number of bowel movements and is not very painful, then you know this is not a chronic problem. However, if you cannot get it under control, it is best to see your doctor immediately. Many of my clients, especially Body Type Two — those drawn to spicy and fatty foods — create an imbalance through the overproduction of bile. When bile is thin, or frequently produced, it creates peristalsis (muscle action of the colon). This causes several bowel movements in rapid succession, or diarrhea. A stressed gallbladder can create the same problem, also directly related to the type of foods you eat.

If foods are digested correctly and the body is able to get from them the nutrients it needs, you would never have diarrhea. As our health improves, our stools become more like they were when we were babies or young children; they are rather formless. If you eat two to three meals per day, then you should have two bowel movements per day in order to stay healthy. Some of my clients have actually told me they have diarrhea because they have more than one bowel movement in twenty-four hours!

There are those body types that fluctuate between constipation and diarrhea. Poor eating habits can cause an endless cycle of several days of constipation followed by several more of diarrhea. It often becomes very troublesome and persistent. In this case, the results of treating the problem with enzyme supplements have been phenomenal. In most instances, the diarrhea becomes a thing of the past. On the other hand, if the diarrhea is infrequent, it may not be directly caused by what you eat, but because you cannot digest correctly. Use of the supportive herbs described in the Gastritis Relief section can bring a calm, balanced effect upon the gastrointestinal system. Lipase is the best enzyme to use for diarrhea control, as it works in harmony with the gallbladder. In addition, it breaks down fats. Incomplete fat digestion will lead to diarrhea in certain instances. Amylase will eliminate any inflammation that may occur.

Constipation

My past years of practice as a colonic therapist taught me the importance of a healthy bowel. Colon irrigation confirmed my belief that very few people experience good digestion, assimilation, utilization, and elimination. When digestion improves, so does elimination. Elimination is every bit as important as digestion and assimilation. Toxemia, the outcome of accumulated and unexpelled feces and other waste matter in the body, is a dangerous and inevitable effect of constipation. A clean, clear colon is vital to our good health. Few of us recognize that failure to effectively eliminate waste products causes a tremendous amount of fermentation and putrefaction in the colon. Neglect of this accumulated debris can, and frequently does, result in degenerative diseases.

If you eat processed, fried and cooked foods, starches, sugar, and excessive amounts of salt, your colon cannot possibly stay clear. Even if you have the requisite two or three daily bowel movements it remains clogged with old waste. These foods, instead of nourishing the muscles, tissue, nerves, and cells of the colon walls, actually starve it. A starved colon may let a good deal of fecal matter pass through it, but it is unable to activate the final stages of digestion and nourishment for which it is intended.

When the mineral elements that compose the foods we eat are saturated with oil or grease, the digestive organs cannot process them efficiently. They are passed out of the small intestine into the colon in the form of debris. In addition, the body disposes of a great deal of used cell and tissue matter in the colon. As these undigested waste materials pass through the system without assimilation, they leave a coating of slime on the inner walls of the colon, like plaster on a wall. With the passage of time, the coating will gradually increase in thickness, until there is only a narrow opening in the center. The matter to be evacuated contains much undigested food, from which the body can derive little or no benefit.

Sickness at any age is the direct result of loading up the body with foods that have no vitality; that are not digestible. Allowing the intestines to remain packed with waste matter is further cause of illness.

The Body's Sewage Disposal Plant: The Colon

The colon is a natural breeding ground for bacteria, of which there are two types. These are the healthy, scavenging type such as bacilli coli, and the pathogenic or disease-producing kind. Healthy bacteria in the colon neutralize, dissipate, and prevent toxicity from developing. When too much fermentation and putrefaction occur, the pathogenic bacteria proliferate and ailments result. A normal colon is equipped with a very efficient elimination system.

In simple words, the colon is the sewage system of the body. If we do not keep it as healthy and free of disease as possible, our major organs and glands are seriously compromised. The colon affects every one of the body's organs. It is intimately related to every cell in the body. The primary purpose of the colon is as an organ of elimination. It collects all terminative and putrefactive toxic waste from every part of the anatomy. Through peristalsis, the contraction of the colon muscles, all solid and semi-solid waste is removed.

The colon is the sewage system of the body. If we do not keep it as healthy and free of disease as possible, our major organs and glands are seriously compromised.

The best of diets can be no better than the worst if the sewage system is clogged with a collection of waste. It is impossible when we eat two or three meals a day not to have a buildup of residue in the form of undigested food particles as well as the end product of digested food. The millions of cells and tissues that have served their purpose and are being replaced by new ones become dead proteins of a highly toxic nature if they are not eliminated from the system.

Remember, the colon plays an extremely important role in the final stages of digesting food and energizing all systems of the body. The residue coming from the small intestine (chyme) still contains some uncollected nutritional matter. The colon is supposed to extract this matter. It mulches this matter and then transfers the liquid and other elements through its walls and into the blood stream. The blood carries it to the liver for processing. Obviously, if the feces in the colon have putrefied and fermented, any nutrients present would pass into the blood as polluted products. This is a condition known as urea. This substance can be measured through a urine test to determine the level of toxicity of the body.

Lumen of the colon contains intestinal flora for lubrication. Fecal encrustation interferes with the infusion of this flora. It also causes a severe lessening in the peristaltic action needed for evacuation. Further, the ability to absorb and use the additional nutrients present in the waste residue which enters from the small intestine is badly compromised. The human bowel can extend to five times its normal size. Imagine the quantity of accumulated, packed feces that can be stored in an unhealthy colon! All this fecal matter in the bowel only serves to intensify a state of constipation.

Anyone who eats three meals per day but experiences less than two bowel movements in that same time frame is in a constipated state. It is a natural function of our bodies to eliminate waste after digestion. Constipation is self-poisoning—brought on by lifestyle, stress, too many sugars, and poorly digested food. Auto-intoxication; (poisoning by toxic substances generated within the body), begins with an inability to digest food. This in turn creates toxic fecal matter that remains in the colon. These toxins soon begin to be reabsorbed into the bloodstream. See Appendix: I for more information.

There are four ways in which the body rids itself of wastes. The colon, kidneys, lungs, and skin are the four major organs of elimination. Up to thirty-six different toxins can be produced in the bowel, to be released later into our system. These toxins affect every cell in our bodies. To assist with elimination, enzymes are absolutely necessary. For those who suffer from chronic constipation, there are enzymes that should be taken after meals and before bed. Others can restore vital, healthy bacteria in the large intestine. These bacterium-fortified enzymes act as a natural antibiotic in the intestinal tract. Most overweight people are constipated. This is a topic that will be dealt with more extensively in Chapter 10.

Normally, *digestive enzyme formulations are taken at the beginning of meals so that they arrive in the stomach at the same time as our food.* I have heard some people say they have no problems with digesting any kind of food, yet they may have difficulty with regular elimination. I believe people say these things because they are not well informed. It is the purpose of this book to educate you about the importance of enzymes. Only then can you make wise choices for the maintenance of your health.

Acid/Alkaline Balance

Our bodies naturally seek a point of balance in all our systems, and the process of digestion is one of the most important ones. Good digestion will affect your breathing, elimination, nervous system, blood pressure, insulin/sugar controls, and hormones, among other functions.

The body works continuously twenty-four hours a day to maintain the proper acid/alkaline balance in the body. This is known as homeostasis. If this balance is not maintained, the metabolic enzymes and all other biochemical activity will not function normally.

Acidosis is brought on by acid-producing foods, the greatest of which is sugar. It can be controlled by the body's two buffering systems, the lungs and kidneys. These systems control the fluids composed of sodium, protein, and phosphate. The vagus nerve, running throughout the entire body, plays a vital role in good health. If it is pinched by a hiatal hernia, the hydrochloric acid level is reduced. When food is prevented from entering the stomach because of hiatal hernia, it is not broken down to the extent it should be because hydrochloric acid was not released. When food reaches the stomach, bile, an alkaline, is produced. This acts as an abrasive soap that washes away the mucus coating the stomach uses to protect itself from extremes of alkalinity or acidity. The results are putrefaction or fermentation of food, rather than digestion. Hydrochloric acid is the only acid beneficial to the body and is necessary to digest protein. I do not recommend taking this acid in synthetic form, because I believe it is important for the body to manufacture its own. Enzymes assist in that process. If your pH balance is abnormal, you will be forced to breathe very quickly, to counteract the acid imbalance. Because the vagus nerve runs near the heart, the pulse and blood

Good digestion will affect your breathing, elimination, nervous system, blood pressure, insulin/sugar controls, and hormones, among other functions.

pressure increase. Added to this, the stomach becomes irritated and spastic. The liver and gallbladder overproduce bile, further irritating the digestive tract. Then, the pancreas will produce too much insulin, leading to low blood sugar. Within the small intestine, a reduction in alkalines occurs, causing the whole system to become excessively acid. The large intestine responds with either diarrhea or constipation! Conversely, alkalosis is brought on by too many antacids, diuretics, and potassium depletion. In a situation of excess alkaline, the pineal gland overcompensates by producing too much hydrochloric acid. As in all things in the natural world, a perfect order of balance is always maintained. That is called homeostasis, and it applies to our digestive systems as much as it does to the path of our galaxy through space.

What starts this domino effect of distress? Most people become upset about something in their lives at least several times a day. Are you aware that your stomach contracts at even the smallest negative feeling? Recall some times when you have heard bad news, and it felt as if you had been hit in your stomach, or you grab your stomach. Feelings lead to biochemical changes in the body. Emotional ups and downs trigger these as much as biological ones do.

We know that when the body cannot get the fuel it needs to run itself efficiently, it will look to itself for food. For example, when it needs calcium and cannot get it because of poor diet, it will take calcium from its own bone. Added to this, the endocrine glands, your body mind, become sluggish. These glands are crucial to good health; they control the immune and nervous systems. Without the endocrines functioning properly, imbalances occur in our sugar levels, brain food, and hormones. We can experience erratic premenstrual syndrome, bowel problems, fatigue, and a depressed immune system.

Enough has been said about imbalance in digestion. The big question now is, *What can I do to prevent this from occurring? If it already has, what do I do?*

Laboratory tests and the Biological Terrain Assessment ™ using blood, urine, and saliva have proven time and time again that plant enzymes support correct pH balance. I have taught and used the test to support this. I ask my clients to ingest plant enzymes with meals. From a urine specimen, I have observed a constant pH balance of 6.2 to 6.7, with some variation expected due to body type. The optimum pH in urine is 6.4. Doctors who use enzyme therapy have documented the same results thousands of times.

Case History

A very beautiful young woman, a former client of mine whom I had not seen for several years, came to my office recently. The last time we were together she had just been married and was expecting her first child. She was particularly lovely that day she told me she was pregnant. As the health of her child was very important to her, her OB/GYN doctor sent her to me for advice on diet and enzymes. Throughout her pregnancy and the birth she carefully followed her enzyme program.

However, after the child was born, she became negligent in her daily routine. I have seen this happen so many times. Because she felt better, she did not think she needed the enzymes any more, and stopped taking them. Even though all symptoms are removed if you stick with the daily regimen of enzymes, you still need to continue your program. This young woman enjoyed good health for awhile after she stopped using them. But soon, the old habits and cravings for sweets, coffee and pizza were back. She once again experienced fatigue, skin breakouts, hiatal hernia symptoms and gallbladder nausea, brought on by digestive problems. Her husband lost his job, and she had to return to work, leaving her child in someone else's care.

Although I was sympathetic to her plight, I simply suggested that she treat herself with the same love and care she had shown for herself and her baby during her pregnancy. Knowing that she could not change her body type because it is something we bring with us into the world, we worked on an appropriate diet for that type and added the enzymes. Three days later, she called to cancel her appointment. She was already back in control and knew that she would continue on the path toward good health. She learned that her good health was something she owed to herself and her family. To this day I am amazed when I see her. She exudes youthfulness and happiness. People compliment her on the beauty of her skin, and they constantly comment on how she looks younger every time they see her!

Those of us who are in enzyme nutrition hear this story over and over. I cannot say it often enough: Education and supplementation are of extreme importance. **When we understand why we have pain or feel ill, we can make choices and stay in control of our lives**. For so many years we thought that if we had a particular symptom, all we had to do was treat it, make it go away, and then resume our old lifestyle. That is how conventional medicine is practiced. However, when you find a health maintenance program that works for you, you must commit to follow it for the rest of your life. Instead of treating what you have as a disease or its symptoms, you increase the love you have for yourself and others. That love must include the gift of good health to yourself.

How to Overcome Dairy Intolerance

When your body can no longer make the correct amount of enzymes to digest a particular food properly, but that food is still consumed, an intolerance for it is created. From then on, we experience an allergy or adverse reaction to it. It is a toxic substance for us. This is what occurs for people who develop a problem with dairy products. Lactose is the offender contained in dairy. The enzyme lactase breaks down lactose. I have my clients take lactase along with other enzymes, although it is not the most important one. Protease is most important in the digestion of dairy because it is a protein itself. Lipase must be included because of the fat content in dairy products. One should take a multi-digestive formula rather than a single lactase product.

My advice to those who suffer dairy intolerance, or any other kind of food allergy, is to take the enzyme supplements for at least twenty-one days before trying to eat the food you are allergic to. It takes that long for the system to make a positive response to the newly introduced enzyme support. Then, be sure to take digestive enzymes each time you consume those products. I strongly encourage you to follow this program before you reintroduce foods with which you have had problems in the past.

Sugar Intolerance

I believe that most of America is glucose intolerant. By that I mean we have become very lackadaisical about how much food we eat containing high levels of sugar. Sugar is used as a preservative in most foods. The average person simply is unaware of this fact. It appears in surprising places. Any food that has been processed, canned or frozen includes sugar. We give our children an amazing amount of sugar without realizing it. We are a nation of sugar addicts. A whopping 92 percent of all foods sold in America is combined with sugar to a greater or lesser degree. It is hidden in most food products on the market. The average American eats one cup of sugar every day, or 210 pounds a year! Any label that claims "no fat" or "low fat" has high levels of sugar listed on it. Certain soft drinks intended for young people contain sixteen

The average American eats one cup of sugar every day, or 210 pounds a year! Any label that claims "no fat" or "low fat" has high levels of sugar listed on it.

to eighteen teaspoons of sugar in twelve ounces of liquid! Cereals sometimes contain more than sixty teaspoons of sugar!

Our glucose balance is the most precious and high-precision balance in the body. The brain needs one hundred forty grams of glucose, or we will get a headache. Added to that, we need an additional forty grams to make red blood cells. One hundred eighty grams of glucose per day is the requirement for normal function of the brain and healthy blood cells. Yet, we ingest one thousand grams of sugar each day. How does our body handle the surplus? We produce more insulin, which carries glucose to our cells.

Hypoglycemia is the end result of overeating sugar. This happens when the body is forced to make enough glycogen to balance the increased quantities of glucose and insulin in the system.

America is virtually falling asleep because of eating so many simple carbohydrates (sugar). Sugar robs the system of nutrients as well as creating an acidic state. So many children are unable to absorb their lessons in our schools, because their brains are blocked by too much sugar. Fatigue is rampant among adults as well. I have concluded that sugar is responsible for most of the colon problems we suffer. Most of us still do not realize that overdoses of sugar have far-reaching effects, causing numerous health complications.

Enzymes can make a big difference. Amylase and sucrase work well in the breakdown of sugar and help with gaining control of how much we ingest. Protease provides a nontoxic environment. This is one way to start the "cleanup" process. However, each of us must take responsibility for our own consumption of sugar. Remember the fatigue, anxiety and headaches you have while eating a diet heavy with sugar, and it will assist you in your efforts to control your intake.

Diabetes

The medical records of the nineteenth century are very sparse and incomplete when it comes to any cases resembling diabetes. Why? Because type II diabetes, or what was formerly known as adult-onset or noninsulin-dependent diabetes is the direct result of sugar consumption. Refined sugar was not widely used until after Napoleon built sugar factories in Europe in the nineteenth century. In 1850 the average Englishman consumed only seven and a half pounds of sugar per year. By 1976, the average American consumed one hundred twenty pounds. About fifteen million citizens of the United States have type II diabetes. Some people are born with type I diabetes. From birth, the pancreas is unable to produce the necessary insulin. Type two is characterized by an overabundance of it. Insulin is the pancreatic hormone that regulates the transport of glucose—blood sugar—into the cells.

In type two, the body develops a resistance to the action of insulin. When the body fails to respond to insulin, the blood glucose level rises and pancreatic cells pour out more of the hormone in a vain attempt to process it. Eventually, these cells wear out, causing insulin shortage.

Type two diabetes and its complications account for as much as 15 percent of current healthcare costs. Some common occurrences that are suffered by those with diabetes are blindness, kidney failure, and loss of limb. Each year twenty-four thousand people become legally blind, fifty-six thousand develop kidney failure, and fifty-four thousand lose a leg through amputation. At all ages, people with diabetes have twice the normal death rate.

Let us compare diabetes to hypoglycemia. Diabetes includes hypoglycemia in its definition. The unbalanced relationship between blood sugar and insulin is a description of both ailments. It is my belief that diabetes begins first with hypoglycemia, a Greek word that means "low blood sugar." This can be misleading, because it starts with excessive amounts of sugar in the blood. Many people have been suffering from hypoglycemia years before their diabetes is diagnosed. It is so common in the United States that most Americans suffer from it without knowing it.

When we consume large amounts of sugar, the body counteracts with the release of high levels of insulin. Remember, insulin is a hormone secreted by the pancreas. It transports glucose, a form of sugar useful to the body, through the bloodstream and into the cells, where it is converted to glycogen. Glycogen is stored carbohydrates. If it is present in excessive amounts, it is further modified into a fat called triglyceride. Copious amounts of insulin contribute to obesity because of this conversion of glucose into triglycerides. When we consume any food that breaks down into sugar, such as fruits, vegetables, or starches, we are taking in a considerable amount of sugar. Even refined carbohydrates eventually break down into glucose. Sooner or later, your body overreacts. Instead of producing the exact amount of insulin necessary to handle the quantity of sugar ingested, it makes more. This excess insulin is what forces the blood sugar to an even lower level. The consequence of consuming large amounts of sugar causes the body to end up with less blood sugar than it started with, and the result is exhaustion. So what do we do? We eat more sugar for that so-called "mid-afternoon pick-me-up." The body again secretes more insulin, and the cycle is perpetuated.

The widely accepted "Food Pyramid" tells us that we can safely eat refined carbohydrates, but how many of us know what refined carbohydrates are or if they are truly good for us? They are foods such as white flour, cornstarch or white rice. The enzymes once present have been destroyed by the refining process. These refined carbohydrates are processed by the body the same as any other food, and yet they contain no vitamins or minerals. If we eat them, we must draw upon our own store of nutrients normally used in metabolism, thus depleting them. Some people say they do not eat refined sugar, but if you eat flour, cornstarch, white rice, potato starch, bread, crackers, or pasta — you eat sugar! Milk, yogurt, and other dairy products contain sugar in the form of lactose. In fruit and juices, the sugar is glucose and fructose. Learn to recognize these different forms of sugar when you are reading labels at the supermarket.

The enzymes used to correct sugar-related problems are lactase, maltase, sucrase, and the most important one, amylase. It is so important to understand that all carbohydrates will cause you to release more glucose into your blood than any other food group. This will require additional quantities of insulin. Because diabetes is rampant in America, the medical profession created another category of diabetics. Low-grade or borderline cases have been reclassified. The condition of patients in this group is now referred to as impaired carbohydrate tolerance.

A leading risk factor for developing type two diabetes is obesity, especially abdominal obesity. This is due to insulin resistance, or hyperinsulin. These, in turn, are all risk factors for heart disease. Virtually all obese people and those with adult-onset diabetes have hyperinsulin. At least half of those who suffer from obesity are diabetic, and over 80 percent of adult-onset diabetes cases were at one time overweight.

Hypoglycemia

Hypoglycemia is extremely common. It afflicts the majority of Americans. Its symptoms include headaches, low energy level, mood swings, alertness, irritability, hunger, tension, and nervousness. Depending upon what food has been ingested, the symptoms can change from hour to hour. Often, hypoglycemia completely controls a patient who is suffering these symptoms, even though he is unaware of it. Many conventional doctors believe there is no such thing as hypoglycemia, categorizing it among those "nonexistent" maladies like PMS or Chronic Fatigue Syndrome. They refuse to admit to it, yet over and over they see many patients with the symptoms described above. These physicians do not know how to treat them. The hypoglycemia I refer to and treat with enzymes is not a disease. It is a condition that affects a huge percentage of the population. Most people, if asked, would say they have some of these symptoms. We blindly accept this state of health as a way of life. Remember, low blood sugar is not a diagnosis or a disease—it is a response!

What can we do about hypoglycemia? I advise those who come to my office that education is very important. We must learn to identify the amount of sugar we use, especially those hidden ones. Most of us do not even know they exist. The next step is to replace the body's depleted supply of amylase. Along with the enzymes there are certain herbs that assist with the insulin response. Please do not feel that if you have this condition there is no hope. A change of diet with enzyme supplements can do wonders for you. Hypoglycemia involves allergy, or substance intolerance. It is very difficult to determine all the different reactions we have to foods. When we discover what they are, it's even more difficult to eliminate them for the rest of our lives. Therefore, as a preventative, I suggest the use of enzymes to aid digestion and stop toxic reactions.

Symptoms of hypoglycemia can be tolerated for brief periods of time. However, if the blood sugar level remains very low for a prolonged period, then fatigue and dysfunction develop. These can include confusion, hallucinations, convulsions and even lapsing into a coma. The nervous system is deprived of glucose necessary for metabolic activity. Another normal response to hypoglycemia is a significant increase of epinephrine, a hormone secreted by the adrenal glands. More serious symptoms develop such as increased pulse rate, tachycardia (elevated blood pressure), sweating, and anxiety.

Treatment would depend on the primary cause of hypoglycemia. If it is a medical problem that stems from an endocrine or liver disease resulting in the decreased secretion of glucose, then special dietary changes are in order. These changes are aimed at avoiding extremes in the blood glucose level by maintaining a constant one at all times. This diet is high in protein and fat but low in carbohydrates. It calls for frequent, small meals during the day, and one before going to bed.

Another cause of hypoglycemia is cell damage to the liver. If someone is fasting, the liver is the primary source of glucose to the bloodstream. This can also result in an ability to convert glycogen into glucose. Glycogen is actually formed by and stored in the liver and, to a lesser extent, in the muscles. It is changed into glucose and sent to various parts of the body as needed. It's been called the animal starch.

Many people have been told, or assumed, that they have hypoglycemia. Our poor eating habits generate it. Another cause may be an inefficient endocrine system or a tumor. Hypoglycemia is recognized and discussed by most people. The majority of cases, however, are simply glucose intolerance, rather than a true hypoglycemia. In either case, there must be a change in the diet.

The diet must include protein and careful control over the amount of sugars we eat. Some prescription drugs lower blood sugar levels and interfere with the production of insulin. These drugs create a hypoglycemic effect. Hypoglycemia is touted as one of the disorders of the modern age. Because of that, many people tend to dismiss it. In truth, it is more serious than we think.

One of the first things we do for hypoglycemia in the world of enzyme therapy is to fortify the liver. The best choice for this is protease. I always suggest to my clients who have any form of glucose intolerance to make an effort to identify the foods they crave or eat in the greatest amounts. Then, eat several small, low-sugar meals a day. This way, problem foods that should be avoided can be determined. Refer to the Body Type section to find your type, and the appropriate foods for you. Above all, good digestion is essential. Protease and amylase in the right proportion must be used as supplements.

I see clients in person at my office. In addition, I receive calls from people all over the country who ask about body typing and enzyme therapy. I have developed a questionnaire for all of those who seek my help. I utilize this as a tool to assist me in evaluating individual nutritional needs. One of the disorders listed is hypoglycemia. Most people mark it as a symptom they have experienced. I chose "hypoglycemia" over "glucose intolerance" because most people still do not know what that means. But if you ask an average American what hypoglycemia is, he or she will answer that it is the fatigue felt after eating sugar. This is an example of how trendy hypoglycemia has become in the last few decades.

Case History

This brings to mind a child who came to my office one day. His mother was very concerned about his lethargy; he had become a couch potato. He had no interest in sports or school. All he did was watch TV. But even that did not keep him focused. His parents thought he had a learning disability. They put him through a long series of tests at one of the local children's hospitals, only to find he had no learning difficulties. His inability to concentrate, the dizziness and shaking he was experiencing, were still unexplained by the time they came to see me. After reviewing the boy's diet, it did not take long to realize that his system was completely out of balance due to eating far too many simple carbohydrates. He was suffering from all the symptoms of hypoglycemia, or sugar intolerance.

He was willing to eat all the foods we suggested, at the proper times, and to take enzymes. I agreed to drop out of the situation if none of it worked. I made an absolute agreement with him; he promised to work with me. I designed a plan for him that started with enzymes first thing in the morning, between meals and with meals. I explained everything to him, so he understood how the enzymes fortified his system. He and his family saw a real difference within three days. This child told me that he actually felt better within twenty-four hours. He had a complete reversal, and now he is very involved in school work and neighborhood activities. Once we got things under control, I could show his parents why all the vitamins and minerals they gave their son and the methods they used to stop him from eating sugar were ineffective. He suffered an inability to properly digest his food and utilize it as fuel for his body. No amount of vitamins or changes in his habits would work!

Many parents ask me why they should have to keep sweets from themselves and the rest of the family just because their children cannot control how much sugar they eat. Recently a mother complained to me that her child was sneaking sweets into his room. She could not understand why she and her husband were having behavior problems with him. When she asked the child what they could do to help him, he answered, "Just don't buy those things, Mom."

Her comment to him was, "Why should the rest of the family suffer just because you have a problem with sweets?" I pointed out to her that a child learns from parents and other adults. In order for a child to know how to control his sugar intake, the parents must first show that they can control theirs. It is inconsiderate to have something in the house that constantly reminds the child of his forbidden foods. By watching the rest of the family indulging in them, the child becomes confused. He cannot help thinking, "If it's okay for them, why can't I have it, too?" or "Why am I being punished?" That child was very intuitive and absolutely correct with his response. It was obvious to both his mother and me that she too had a problem with sugar. She was not willing to give up her own sugar cravings. Otherwise, she would have happily honored her son's request.

Many times in all our lives, we can readily see other people's destructive behaviors without seeing our own. If you feel fatigued, dizzy, lacking in stamina; if you walk into a room only to forget why you went there, if you are speaking to someone and in mid-sentence you forget what you were saying, it is time for you to look at your sugar intake. I could relate many stories to you, but they are all very similar to those I have told. We do not like to give up that sugar. We think we feel better when we have it. We do not like to deny ourselves; yet America is asleep because we are eating so much sugar. It is all too easy to justify our behavior by saying we have hypoglycemia, and therefore cannot control our sugar cravings. The truth is we do not know how to push away those desserts offered to us! In my own experience, by taking plant enzymes for sugar digestion, the cravings have lessened because I am allowing my system to heal. Try a program of amylase, lactase, maltase, and sucrase and see what happens.

Detoxify and fortify: Purifying your blood. . . Lowering your cholesterol

Detoxify means to counteract or destroy toxins or to remove their effects from the system. In the past, these two processes were employed in a linear fashion. First, through the use of some detoxification process like fasting, we would flush toxins into the blood stream. This caused flu-like symptoms; fever, fatigue and overall aching. Then we would fortify the body with certain nutrients to restore the balance. Now with enzymes, we can detoxify and fortify at the same time without the side effects. Chapter Three will explain exactly how enzymes can be used to cleanse the body and boost good health simultaneously. Enzyme cleansing eliminates the discomforts associated with detoxification while increasing a sense of well-being and higher energy levels. They are a beautiful and non-invasive way to cleanse and nourish.

By now you know how strong digestion, combined with a tough immune system, contributes to a healthy, vital blood supply. I have covered many of the reasons why these systems are often below par, allowing the blood to become weak and polluted. It is important to know that germs and viruses only thrive on decaying matter (toxins). They do not attack clean, toxin-free blood, tissues or organs. Therefore, if we reduce the number of contaminants in our blood, we automatically reduce the number of germs and viruses. There are two ways we can accomplish this. We can depend on our body's ability to cure itself, and we can offer a little assistance to get the job done.

Our bodies are true marvels of creation. Hundreds of interconnected organs and systems are programmed to respond to an endless variety of circumstances

—good and bad. For example, four major command centers are constantly at work simultaneously, removing toxins from our blood. You will discover in Chapter Seven that our bowels, kidneys, lungs, and skin (yes, skin!) consider themselves protectors of our blood, among other things. They work tirelessly at keeping the blood free of unwanted debris, each in its own unique way. Unfortunately, when our bodies are short on enzymes, these wonder-workers are not always effective because of the stress we impose on our systems.

As we get older, our bodies simply cannot effectively detoxify our blood and lymphatic system without help. The lymphatic system filters out waste products. For example, white blood cells attack bacteria. As part of the process, they die along with the bacteria. The lymphatic system ushers both these waste products to the lymph glands where the fragments break down and assimilate harmlessly into the body. For decades, millions of concerned consumers have tried a variety of self-administered ways to detoxify. They experimented with everything—macrobiotic and other vegetarian diets, fasting and herbal regimens. Most of these approaches work to some degree, especially the herbal treatments. However, my studies have convinced me that enzyme supplements are the missing ingredient. When taken consistently, they purify the blood by breaking down its undigested proteins, cellular debris, and other toxins. With the blood healthy once again, the body gradually rebuilds itself and replenishes its storehouse of enzymes. (Remember Dr. Howell's bank-account analogy?) The ultimate result: A balanced body capable of functioning at peak efficiency.

Enzyme supplements are the missing ingredient. When taken consistently, they purify the blood by breaking down its undigested proteins, cellular debris, and other toxins. With the blood healthy once again, the body gradually rebuilds itself and replenishes its storehouse of enzymes.

I find it interesting that when dealing with people I know, either friends or family, it is very difficult to advise them about their nutritional well-being. It is amusing to learn that you simply cannot tell people whom you know what to do! I am addressing this because I have suggested the use of enzymes and dietary changes to my own family and close friends. In keeping with our human nature, they all wait until they have manifested a condition or illness before they make any changes. When they finally do take action, it is invariably the result of what they have learned from someone else's book or work. My word to the wise is, if you have found the answers to a loved one's health problems in this book, they may not be ready to listen to your pointing it out to them. Be comfortable with what you have learned—the best example for your friends or family is to see you working diligently at staying healthy.

When family and friends do rid their diets of sugar, the results that manifest are so encouraging. Because our nation has such a sweet tooth, we have become so enzyme-deficient that we cannot recognize the battle raging within our bodies. The symptoms are so numerous now that there is a great deal of crossover into other maladies besides sugar addiction. Fatigue, aches, pains, depression and inflammation describe a multitude of disorders. One of my relatives recently read a book about Crohn's disease. He is having a problem with his colon and the book detailed the number of malnutrition dilemmas we are facing because our metabolic enzyme output is so badly hindered. The digestive system is so compromised that this colon disorder has now manifested as Crohn's disease. My loved one eliminated almost all the simple carbohydrates, desserts, breads and some of the pasta from his diet. It worked for him and now he feels so much better. It does not matter whether he heard it from another source or me, the important thing is he understands what he had been doing to himself, and he has taken steps to change his lifestyle. I reiterate, I cannot emphasize too strongly the importance of eliminating sugar from your diet. Get rid of those sweets, those white breads and tortillas and start giving enzymes to your body. I am speaking to myself just as much as to anyone else.

Your metabolic enzymes are inhibited by eating foods that create residue in the small intestine. This waste matter keeps the villi from absorbing nutrients. The system can no longer depend on the metabolic enzymes, which have been rendered useless. They are unable to respond if called upon, because of the toxic conditions we have created by eating the wrong foods day after day, year after year.

The protocol for a very toxic colon that can involve explosive diarrhea calls for the return of a good pH balance. It is almost impossible to achieve if you are nutrient-deficient. If your small and large intestines have bacteria or fungal-form takeover, the best recourse is to use ***L. Plantarium,*** a natural bacterium. (See page 210.) I then call in my protease, the big detoxifier. Protease is a scavenger antioxidant. Its miraculous properties cannot be praised enough. Once in a while, you may find someone whose gastrointestinal system is so irritated that taking high doses of protease may cause a burning sensation. If that is the case simply take your protease with a meal or small amount of food instead of an empty stomach until healing of the area has taken place. These are the open lesions we have referred to earlier in this chapter under Ulcers. Our job is to calm down and repair that system. The bacterium, *L. Plantarium,* will not create the burning. We will restore the natural pH and the "good" bacteria. When this takes place, it works like a natural antibiotic because our body is able to make its own reserves. The next best enzymes for this are high amounts of amylase and lipase, with sufficient quantities of cellulase. Used with herbs such as marshmallow root, the mucosal lining of the intestinal tract will be replaced and the tract soothed. Combined with a small amount of Gotu Kola, the marshmallow root creates balance. When you use more than one herb, the healing scenario becomes more complex. Adding slippery elm creates a repair system. This is the kind of formulation I would use until everything is back to normal.

To be toxic means we have self-poisoned our bodies either with undigested food or through autointoxication. This means we have poisoned our system through toxins locked in the blood stream from something we have not digested properly. Toxins are trapped in our colon where they leak into the blood and then we have to deal with the adverse effects. There are thirty-six poisons released into the blood from undigested foods. These poisons, or toxins, are substances that have been ingested, inhaled, absorbed, or have developed within the body. Even in relatively small amounts they can cause structural damage or functional disturbances.

Cholesterol . . . Is it really the bad guy?

We have known for decades that over time, a high fat diet will raise the cholesterol level in the blood and increase risk of a heart attack. We have also known that unless you are a smoker or have a genetic defect, you have the ability to reduce your cholesterol level. More recently, we have learned that when you reduce your cholesterol level, you also reduce the amount of high-density lipoprotein in the blood. This is not ideal, because HDL plays a very important role. It drives excess cholesterol from the bloodstream into the liver which removes it from the body. Fortunately, regular exercise offsets the loss of HDL good fats caused by low-fat diets. We often overlook the fact that the body manufactures its own fats. Although we need a certain amount of fat to maintain good health, it is all too easy to accumulate harmful levels. Knowing our individual body type and the foods we should eat is very helpful in preventing our system from overproducing our supply of bad fats.

Cholesterol is essential to good health because it is a major building block for all the body's hormones.

In spite of our best intentions, however, a majority of us have too many undigested fats floating in our blood. It is very rare for me to find fat-free blood when I view samples under the microscope. The presence of excess fats in the blood is the result of too much fat intake or fat production, too little exercise, or a deficiency of the digestive enzyme lipase, which breaks down fats. Nearly everyone whose blood samples I have evaluated are lipase-deficient. Taking lipase enzyme supplements with every meal easily remedies this.

As a clinical nutritionist, I have no doubt in my mind that good nutrition is the best method of prevention. The general public, because it is uneducated when it comes to good nutrition, views this way of life as too simplistic. For example, the press constantly feeds us information from doctors and government health reports. Because we all pay such close attention to the media, most of us believe that cholesterol causes heart disease, and that we will prevent heart disease by avoiding cholesterol. I believe that no true scientist genuinely believes it. Even though what we hear is just opinion, we have been left with the idea that cholesterol is bad for us.

Is cholesterol bad or is it necessary? Our bodies manufacture fifteen hundred milligrams of cholesterol per day. If we do not eat foods containing it, our livers make it for us. This tells me that cholesterol is important. To preserve that all-important state of homeostasis, I believe cholesterol, in balance with blood sugar, blood pressure, sodium, potassium, electrolytes and, everything else, needs to be present. If very high levels of cholesterol are not good, I suspect that neither are very low levels.

What does cholesterol look like? It is a waxy, fat-like substance. When we look at live blood just after someone has eaten or while they are digesting, we expect to see a certain amount of cholesterol floating around in the plasma. If it is not there, we know a big imbalance exists, one that needs correcting as soon as possible. Cholesterol is essential to good health because it is a major building block. It is a component for all the body's hormones, particularly the sex hormones. It plays a major role in the endocrine system, it helps form the protective myelin sheath around nerve endings and it helps provide the bile salts in digestion.

I have read several books concerning research on cholesterol. I was surprised to find a study done in America with people having blood cholesterol levels of anywhere from 200 to 225 milligrams. They lived longer than those with levels of 150 to 200 milligrams. That makes a statement that differs greatly from what the media tells us. I have consulted some of my medical doctor colleagues who say that very low cholesterol is associated with some diseases. They are more comfortable when their patients' readings are higher than 150 to 200. We know that a certain amount of cholesterol is very important for the maintenance of good health. However, that amount will vary depending on body type.

It is intriguing to me that no-fat and low-fat foods are all high in sugars. This is a primary example of trading one problem for another. My physician friends tell me they use cholesterol readings as indicators of possible nutrition problems. They are not surprised if cholesterol levels are high in patients with heart disease. Rather, they are concerned that this indicates the causes for both high cholesterol and heart disease are the same. It is sugar consumption more so than fatty foods which drive up cholesterol levels because sugar causes the liver to make large amounts of triglycerides.

In my office, as in those of countless health professionals using enzyme therapy, high or low cholesterol is not a problem. I say that with complete confidence. Whether clients suffer from high cholesterol because of an inherited condition or create high cholesterol through improper diet, they are given a formulation high in lipase, along with other supportive enzymes, and Brindall Berry combined with other herbs. Their high cholesterol level can drop within a twenty-one day period. Laboratory tests, university findings, and our research support this fact. Immune-deficient clients with low cholesterol are also brought into the normal, healthy range with enzyme supplements. For some it happens even more quickly when they make positive dietary changes.

It is important to discuss the research on high-cholesterol foods. The information available to most of us tells us that we should not eat eggs or animal fats because of their high cholesterol content. Even though it has been proven that eggs do not present a cholesterol problem, we are still told not to eat them. It is worthwhile to investigate who the sponsors are of the tests that are run, and how the information is delivered to an unsuspecting public. I feel it is not in our best interests to stop eating eggs, because they are very high in necessary amino acids. We hear that cholesterol will escape into the blood, lining the blood vessels with the harmful plaque that creates heart disease. Many doctors will now tell you that it is not cholesterol that creates plaque as much as it is the presence of free radicals.

I personally advise Body Type One to start the morning with eggs. It amuses me how many people question the cholesterol content in eggs. I show my clients the documented research so they can be comfortable with knowing that the egg is safe to eat. I consider the egg white to be one of the ideal protein sources. Cholesterol is only present in the egg yolk. Clinically, eggs are highly nutritious, because they contain important amino acids. Science uses the egg as a standard measure for all protein. If a health professional advises that eggs are bad for you, ask to read the most recent information about cholesterol research. If you have an allergy to eggs, then of course they could be harmful for you. This could be a warning that perhaps your body is not making the correct amount of protease. When I recommend certain foods to a client, I always make sure the foods are eaten with the right amount of enzyme supplements to assure proper digestion. For me, any concern with cholesterol would begin with taking a look at lipase activity. Next, I would consider the body's ability to handle lipids, and what should be done to balance the body type with foods and enzymes. For more information on cholesterol, see the Chapter 9 on Body Types.

How to Conquer Free Radicals

A free radical is any molecule with an unpaired electron. It can be a molecule of any kind of matter. These molecules are slaves to the laws of physics dictating that a stable state must always prevail. In our bloodstream, they will steal an electron from any atom, without discrimination. This creates another free radical, and another and another . . . soon, a good deal of damage is done and the body's health really begins to falter. Specifically, molecules one of the many areas targeted by free radicals. When we observe live blood under the microscope, we find the platelets clumped together, forming thrombocytes. When the body or artery is under attack, thrombocytes rush to the trouble spot and form clumps. This creates a blood flow problem. In addition, the body naturally makes a fiber-like protein found in the plasma as well. Then, the appropriate minerals are dispatched to the aggravated site. If calcium is sent, it reacts to the area under siege and forms its own type of plaque.

The strongest scavenger antioxidant is protease. This enzyme delivers other antioxidants into the bloodstream.

The questions many doctors ask are "What part does cholesterol play in the creation of plaque? Does it arrive early or late in the procedure? Was it formed because of eating a high-cholesterol food, or was it brought on because junk food seems to encourage the production of free radicals?" Dr. Elmer Cranton, President of the American Holistic Medicine Association and author of *Bypassing Bypass*, was the first to propose a theory that has influenced the thinking of many health professionals. He suggests that perhaps the role of cholesterol is to help protect the body from free-radical oxidation.

Cholesterol is one of the body's natural antioxidants. This means the body manufactures it to guard against the oxidizing damage by free radicals. While sheltering the afflicted area, cholesterol is converted to its oxidized form, which is harmful to the blood cells. The HDLs counteract this effect. LDLs are associated more closely with heart disease than with total cholesterol.

In the nutrition field, many have concluded that cholesterol is not the enemy. High-serum cholesterol (LDL) results from the imbalance in the body. Enzyme antioxidants, along with vitamins found naturally in herbs, are the answer to the body's need for nutrients with which to rebuild itself. The media bombards us with information about popular antioxidants on today's market. Pine bark or grape seed, which I prefer, are two of them. Rose hips are better than synthetic vitamin C. Alfalfa is another product that works well in the restorative process. Siberian ginseng is a nutrient as well as an antioxidant. I use **Ginkgo biloba** leaf and extract regularly. Remember, the strongest scavenger antioxidant is protease. This enzyme delivers other antioxidants into the bloodstream. With a correct formulation of all these enzymes, herbs and nutrients, the invaders do not stand a chance!

Obesity

Over one-third of Americans are overweight. We spend over thirty billion dollars per year on exercise and diet pills, all because we are trying to get rid of the fat cells in our bodies. After we spend all that money, most of our efforts fail, and lost pounds are usually found once again. Visit any newsstand or magazine section of a supermarket, and you will notice that almost every book or magazine features an article on weight loss, a new way to build the body, or a breakthrough diet. All of these books and articles say much of the same thing. There is disagreement regarding low fat, more protein, no protein, or all vegetables. It seems no one takes the time to explain to the public that the greatest imbalance is enzyme imbalance. One form of help for weight loss touted is diet pills, either prescription or over-the-counter. Almost always these pills contain an appetite suppressant that helps you cut back on your food until you can get control of it. Others are made up of medications that block insulin production and deter us from addressing another function of our metabolism. However, when we stop taking the pills the weight always comes back. Most of these medications have side effects, so they can be taken for only short periods of time. Consider the terrible side effects of Fen-Phen.

For some people, a lifelong weight problem means a continuous cycle of gain and loss that puts an incredible amount of stress on our bodies as well as our emotions. Whatever the situation, shedding excess pounds is indeed a difficult task. What is more, the road to a slender body is paved with nutritional dangers. Most diets fall short when it comes to providing adequate nutrition. Some diets suggest certain foods or food groups, while others say to eliminate this food and try that one. The most critical point to remember is that each cell in your body requires forty-five different nutrients to remain healthy! An article published in the *New England Journal of Medicine* in 1991 stated that research has proved both men and women who diet frequently suffer from heart disease more often than others.

To make matters worse, habitual dieting is a setup for weight gain. While you eat less to lose unwanted fat, your body thinks it is starving. It has a natural survival mechanism that allows it to cling to calories or fat, making weight loss even more difficult. If a diet recommends less than 1200 calories a day, it will not provide adequate nutrients. I know you have heard all this before. What you may not know is that people are overweight because they cannot digest food efficiently or receive delivery of the nutrients to their cells. Therefore, the body cries out for more and more food. To answer that call, we eat more without assimilating anything of value, and gain even more weight.

Enzyme supplements added to a meal will help the foods through the digestive process, finally delivering the nutrients. Enzymes taken between meals will assist with cleanup of undigested food particles. Enzymes are critical for the treatment of obesity. Overweight people, unable to control food cravings, continue to assault their bodies and immune systems with high-calorie foods that are devoid of nutritional value. Plant enzymes solve these problems and keep us from doing any further harm to our health.

Below are samples of plant enzyme formulations I would recommend to take at each meal for normal digestion. However, if you are overweight or have a condition you may need 3 to 5 times the amounts shown below.

Amylase: **At least 5,500–12,000 units of activity, because we typically consume so many carbohydrates.**

Lipase: **At least 500–2000 units of activity; for poor fat digestion.**

Protease: **30,000–50,000 units of activity per dosage. Must work in alkaline, acid, and neutral parts of digestive tract. There are many varieties of protease. Best formulation contains all three.**

Cellulase: **400–600 units of activity.**

Lactase: **250–360 units or more. Very important but expensive. For dairy product problems.**

Maltase: **About 200 units. Breaks down maltose and other sugars.**

Sucrase: **80 units minimum. Breaks down common sugars.**

I would suggest this combination if you have any of the needs outlined in this chapter. If you have a particular situation, such as a stressed gallbladder, you may want to take some of the soothing herbs I mentioned in addition to these enzymes.

Enzyme Labeling

There are no regulations governing potency in the labeling of enzymes, which can make it difficult to decide which enzyme product best meets your needs. The accepted standard of labeling from the U.S. Food and Drug Administration is the Food Chemical Codex (FCC). Products labeled using the FCC will have unit measurements expressed as follows.

Amylase: DU (Alpha-amylase Dextrinizing units)
Protease: HUT (Hemoglobin Unit Tyrosine base)
Lipase: LU (Lipase unit)
Cellulase: CU (Cellulase unit)
Lactase: ALU (Lactase unit)
Maltase: DP (degrees Diastatic power)
Invertase: IAU (Invertase Activity unit)

We are used to comparing food products based on weight, but with enzymes it is the activity and potency which we should be looking at. Be wary of products which include fillers or list contents by weight rather than proven activity units.

Chapter 3

Strengthening Your Immune System with Enzymes

Attack the Bad Guys!

Does your immune system really need to be strengthened or rebuilt? Absolutely! Why?

There are dozens of bad guys attacking the very system that is designed to do its own attacking. The main task of the immune system is to destroy toxins and other foreign invaders that take an enormous toll on the body. If you are old enough to read this book, your immune system is old enough to need continuous attention.

Of course, the first thing we all need to do is minimize our intake of bad guys. This means avoiding food and water containing agricultural chemicals, as well as limiting sugar which stresses our pancreas, kidneys and liver. In addition, we should cut back on simple carbohydrates that begin depressing the immune system less than an hour after we eat them. We need to think twice about eating meat, along with all the poisons that existed in the animal before it was killed. Three more items that we cannot omit from our list are stress reduction, regular exercise, and positive thinking *at all times*. I must admit that after writing this paragraph, I was tempted to become depressed, too. Yet if we desire total health, none of these are exaggerations.

If we work hard at the prevention process and eat a nutritional diet, our bodies will produce several million immune cells daily and earn the title "healthy." Unfortunately, the majority of us have many reasons for not practicing prevention. The result is that we put our immune systems on red alert.

As Important as "Balance of All Things" Is to Nature, So It Is to the Health of the Human Body

The common denominator among my clients with serious conditions related to a depressed immune system is an inability to digest food properly. Some of the most common of these disorders are Epstein-Barr Virus, Chronic Fatigue Syndrome, Systemic Candidiasis and HIV infection. When poor digestion occurs, the foods we eat are not broken down into the basic building blocks needed to rebuild cells and generate energy for metabolism. This brings on two big negatives: cell starvation and the release of partially digested food particles into the bloodstream. Starved cells cause malnourished tissues that put the brakes on important chemical processes. This results in less movement of the toxins out of the cells. If that is the case, then fewer nutrients are introduced into the cells. There is less energy production for all cells, including those of the immune system. This reduces the numbers of white blood cells that are critical for fighting infection. Overall cell weakness starves the body of fuel and oxygen. A very high percentage of the population suffers from incomplete digestion, but they have yet to develop serious immune system disorders.

When the body releases partially digested food remnants into the system, it creates at least three major pathological problems.

(1) These remnants become irritants, causing inflammation of the mucosal wall of the intestines that in turn releases powerful and damaging chemicals. These chemicals injure the intestinal wall enough to increase its permeability. The foreign proteins in the undigested food can leak directly into the lymphatic channels of the intestinal wall and gain access to the body's entire circulation system. Here they are considered foreign objects, so the defense mechanism mobilizes an attack by the fighter cells of the immune system. Ultimately, the battle-weary soldiers begin to run out of fuel and fail to reproduce sufficient numbers of white cells.

(2) Food remnants also provide fuel for the overgrowth of fermentative fungal organisms such as Candida albicans and other parasites that further drain the immune system.

(3) The general inflammation brought about by these wandering particles causes metabolic reactions that consume large amounts of oxygen. This produces free radicals that damage cell membranes. Other, more complex damage occurs as well, but the medical terms and explanations are more intimidating and confusing. It is sufficient to say that the combined effects of poor digestion and errant food remnants clearly work together. They bring about a vicious sequence of events that leads to an obvious and progressive weakening of the immune system.

I have had many clients in my office over the years with candida overgrowth and other yeast infections. Even after reading several books on the subject and trying extreme diets, each one of them remains extremely fatigued. They have a desperate look about them. By the time they get to me they have tried so many things and have spent enormous amounts of money. These clients challenge me to help them.

Candida is one of the many fungal forms known to the human body. I have met with several candida support groups and it is amazing to see how much freedom they have gained simply by understanding the real nature of the illness. Candida is natural to the body and only becomes candidiasis (overgrowth) when the immune system is not functioning efficiently. Fatigue and depression are two of the most significant symptoms. Good diet alone cannot control these symptoms. We must evaluate the individual bodies' inability to break down foods that compromise their immune systems.

I do not put anyone on an extreme diet or regimen. I suggest they learn about their body types and which foods sustain them. You will read more on body typing in Chapter Nine. Many people ask me questions about diet. Some ask specifically why others can easily eat mushrooms, but they cannot. It is not because mushroom is a fungus, rather it is the inability to break down the mushroom that is the important factor. The answer can usually be found in one's genes or enzyme potential. A big part of my practice involves working with people who have fungal problems. We must always consider individual body type and how the same food can affect a dozen different people with twelve different reactions. The only food that seems to cause havoc for everyone if eaten in excess is sugar. Within weeks after treatment began, dramatic changes took place in the lives of these clients. I observed a big improvement. Later in this chapter, I will discuss candida overgrowth in more detail.

Relief for Ear Infections

Ear infections are very common among teenagers and younger children. These are the ages at which children tend to eat a great deal of sugar and other junk food. This only worsens their condition. The amount of mucus in their systems increases due to a poor selection of foods. The immune system is depressed. When I complete a Live Cell Analysis on them, they all have fungal forms in common. I suggest they take high dosages of protease between meals to boost the immune system. I encourage all of them to pay close attention to how much sugar and other junk food they are eating. A multiple enzyme product will counteract their poor digestion if taken at meal times. These multiple enzymes must be taken with snacks as well. Children love to snack after school while watching TV. Parents are responsible for keeping the enzymes handy if they want to prevent sluggish immune systems in their children. Doctors involved in enzyme therapy begin babies on enzymes as soon as they are home from the hospital. They use only pharmaceutical grade plant enzymes.

Bruising: A Sign Of Deficiency

Normal bruising is not a concern if it happens from bumping into something or perhaps a fall. These bruises fade with a normal passage of time. There are many people who bruise quite easily, especially as they age. Those bruises tend to linger a long while before they fade. As we age our ability to make digestive enzymes is lessened and this leads to a needy and deficient state of health. In preventive medicine, bruising is always a sign of someone lacking nutrients and at least a slightly depressed immune system. If you fall into this category, I suggest you try the natural digestive enzymes for support in delivering nutrients.

We can take vitamins and minerals in whole food form or synthetic. This does not mean they are always delivered to our cells. I recommend you do some research of your own, and look at the studies showing most people to be nutrient deficient. Many eat the best quality foods and take what they regard as good vitamin supplements. Yet they still bruise easily or catch cold frequently.

Taking vitamins and minerals does not necessarily mean that the nutrient deficiency will be relieved immediately. If they are not the right balance of nutrients for your body type (genetic needs), you won't have the expected results. Find out what caused the problem. Quite likely it will be a case of toxicity in the system. We can cause a toxic reaction by taking improper nutrients. A toxic body is a deficient body. Highly toxic conditions manifest in various ways. It is best to treat toxicity and deficiency at the same time. Before we knew about the miraculous properties of plant enzymes, it was nearly impossible to do so. Now we can assist digestion by cutting back on the cause of the toxic state. We can cleanse and fortify the liver and blood along with feeding and reinforcing the glandular and nervous systems. Within a short period of time, the bruises will fade and will not return. What makes it so rewarding for me is that my clients can see and feel the difference in their health. At the end of this chapter you will find more information on enzyme dosages.

Immune System

Overweight persons have a continual battle to fight between the excess body fat they carry and their immune systems. The weaker their immune systems, the more they are subject to colds, flu, serious infections, viruses, heart disease, diabetes, and degenerative disorders. What is worse, when the immune system is depressed, the body is much more prone to weight gain. This causes an emotionally painful, self-perpetuating cycle. Only one of the four body types, Type Three, can carry extra pounds and not suffer fatigue or a depressed immune system. However, if high blood pressure, constipation or insomnia develops, the Type Three will also contract trouble with the immune system. The final result is that they, too, will be as vulnerable as the other three body types. Please see Chapter Nine for more information on body types.

The Common Cold Should Not Be Common

The common cold is an acute and highly contagious virus infection of the upper respiratory tract. At least twenty identifiable viruses have been found to cause colds and they attack anyone with weakened resistance. It has been my experience that almost all of my clients who take enzymes consistently no longer have the colds that plagued them many times throughout the year. How is this possible? I believe it is because they are finally digesting their food completely, eliminating undigested food particles out of their systems! Those food particles create distress and harbor bacteria that depress our immune systems. **Body types One and Two** tend to be mucus makers because of poor digestion.

If you are looking for total health it is so important for you to educate yourself about all causes of poor health. Be certain you know the facts about the sugars, the nutrient robbers, and mucus makers. If this is not enough to convince you, go back and reread the sections that deal with sugar intolerance.

Say Good-bye to Candidiasis

Candidia, a yeast-like fungus, is one of the many fungal forms that can create problems in the human body. (Known as Candidiasis.) There are thirteen different fungi that can be seen in a Live Blood Analysis. They all share the ability to cause fatigue and many other debilitating conditions. I will not spend much time here on the problem of Candida. Hundreds of books and diets on the subject are on the market. Remember, these diets only work while you are closely following them. The stress suffered from following the diet is almost as great as the candida. To keep it very simple, the fungal forms are proteins. When we take protease between meals to consume the excess radical protein, we have done away with the cause of the Candida. The formulation I recommend is at least 600,000 units of protease activity three to five times a day. In the practice of enzyme nutrition, health professionals sometime give over 3 million units of protease in one day. I suggest the food plan corresponding to your body type (Chapter Nine). In addition, you need the proper enzymes for support of the endocrine system along with the miracle-working protease to digest the spores of the fungal forms.

I believe it is important for you to have more information on the part played by the microorganisms that live in our intestines. They are critical to our good health. We carry more than four hundred species of indigenous bacteria that are constant companions to our intestinal tracts. Our health is maintained by keeping them in balance, so that no one organism becomes dominant among these flora. Normally, this balance is preserved by the resident bacteria. We are subject to both internal and external stresses that can interrupt the balance. If the balance is tipped and an overgrowth of yeast organisms occurs, they overwhelm the normal bacteria in the colon. The most common of these overgrowths is ***Candida albicans***. Candida is only a problem if the overgrowth occurs. A certain amount of this fungus is normal and healthy in the human body.

What does this overgrowth look like and how does it make us feel? Symptoms will vary. Sometimes thrush develops in the mouth. Women often get vaginal yeast infections. Three systems are most profoundly affected by candida: The gastrointestinal, urinary and respiratory tracts. An imbalance in the gastrointestinal tract will cause bloating, poor absorption of vitamins and minerals, food allergies, sensitivity, nausea, and diarrhea. Imbalances in the urinary tract produce frequent urination, pain, burning, swelling, and a strong odor. The respiratory tract becomes susceptible to bronchitis, chronic sinusitis, or postnasal drip.

The most common of the candida symptoms are bloating and the inability to tolerate any kinds of vitamins or nutrients. Candida can directly cause a high incidence of the food intolerance. Many more women have candida because of birth control pills. Those who take frequent or massive dosages of antibiotics, such as tetracycline, are very much prone to candida, because the medication destroys the natural balance of healthy bacteria in the intestines. If you eat too much sugar or simple carbohydrates, you are more susceptible. Women who douche too often will destroy the balance of healthy bacteria in the vagina. Protease is the best of the enzymes for treatment of candida. It rids our systems of the excessive growth of any of the fungal forms without side effects.

As with most illnesses and other health problems, the best treatment for candida is to create a healthy internal environment. Many different products are on the market for our use. The need for **bactocilli** and **bifidus** products are common knowledge to anyone who has ever been formally diagnosed with candida. However, I believe that the new bacteria ***L. Plantarium*** and ***L. Salvarius*** are superior. Different bacteria can be used together to create balance. ***L. Plantarium*** that has two hundred billion organisms, with a growth rate that doubles in less than five minutes. That is a big improvement over the acidophilus we used in the past. Bacteria are preventative and precautionary measures for balancing the system. I favor their use because in the right combination, they balance the pH of the small and large intestines as well. Remember, your small and large intestines are the last places good digestion can occur, so it is critical that proper balance is maintained in them. If not, fungal forms will take over.

Candida albicans does not have a long history. Research shows it came on the scene about the same time we started to eat too much sugar. Another event that contributed to its growth was the advent of widespread use of antibiotics, beginning in the forties and fifties. Those who have the worst symptoms and the most acute emotional depression are the ones who have candida and hypoglycemia simultaneously.

To sum up, the best approach for prevention is to add enzyme supplements to the daily diet, and eliminate sugar. Please be aware of your sugar intake. If you will not cut back on sugar, at least add the enzymes. Some of the food will be digested, some of the nutrients will be delivered, and some of the mess will get cleaned up. To get rid of the invaders, the use of protease between meals will start the correction process. Precautionary measures make a great deal of sense to me. When clients come in for a consultation, I do not ask for radical changes. I make suggestions and give choices.

The best approach for prevention is to add enzyme supplements to the daily diet and eliminate sugar.

However, the bottom line on sugar is . . . it's out! An absolute NO to it. This does not mean all sources of sugar. A piece of raw fruit is good for you, but avoid fruit juices because of the excessive sugar content. It is really quite simple. Read the labels on food products when you are grocery shopping. Choose only those items that are not high in sugar or simple carbohydrates. The benefits you will feel are worth the effort. It is very difficult for Americans to overcome their sweet tooth. I feel it would be far easier to give up the sweets and start to rebuild our systems, rather than suffer the aches, pains, and swelling that come with a strong sugar habit. I have found that when we take large doses of amylase, the swelling and inflammation in the body begin to dissipate right away. The client feels an almost immediate difference.

This takes place after using only one of these miraculous plant enzymes. We do know that amylase with meals and protease in large doses are perfect for those who suffer from candida overgrowth and chronic fatigue syndrome. However, I would never formulate a regimen with just amylase, because I know the importance of the enzymes' working together.

Case History

One of my women clients, a very successful nurse, suffered from candidiasis. After going through the standard medications prescribed for it. She was still very lethargic. She felt she had lost her zest for living. She became so exhausted she had to give up her practice. Her skin broke out and she gained a lot of weight. When she came to my office, I explained how I look at the body's natural processes. From there, I told her we would not fight against her illness, we would simply create an environment where the invaders could no longer live. This, of course, comes through the use of plant enzymes. She began with high dosages of protease between meals, three to five times per day. Along with this, she took the multidigestive enzymes for more efficient digestion. This client learned that she could no longer afford to leave undigested food particles in her system. I also gave her herbs mixed with the enzymes. These can be taken in very small amounts, because the enzymes will deliver them to the cells. The two of them make a very successful health-care team. We paid close attention to her endocrine and immune systems. Within a month, she was back to work. In six weeks, she was completely up to speed. Today, she is still doing beautifully. We have remained in contact through the years. I can call on her for support for those who are desperate and feel there is no way out of the traps they have laid for themselves. It gives us a great sense of freedom to realize we can take control of our health, and we can restore balance to the entire body.

Unhealthy Conditions in the Intestinal Tract Are a Parasite's Dream

A dirty, unhealthy intestinal tract is the ideal environment for parasites. There are at least one hundred thirty-four varieties of parasites that can thrive in the human body. Americans are infested with parasites due to our polluted water system, junk foods, too many cooked foods, and a lack of raw fruits and vegetables in our diets. Researchers in the health care field estimate there are two hundred million people in this country who are infested with intestinal parasites. Once they enter the intestinal tract, parasites have easy access to all parts of the body. They can cause a number of diseases.

Parasites, viruses, fungal forms, and bacteria are protein in composition. Because these offenders are proteins, the enzyme protease works wonders in their destruction. Without any side effects, we can fend off these attackers of our immune system. Here is an example of how protease works. We all know that ear infections are a common affliction in both children and adults. The virus causing the infection has a protein coating that protects the spore, the actual culprit. Protease digests the coating and exposes the spore to the white blood cells waiting to pounce upon it! I think that is an absolute miracle of nature, and in particular the design of the human body.

Americans are infested with parasites due to our polluted water system, too many cooked foods, and a lack of raw fruit and vegetables in our diets.

Parasites occur naturally in the body. They are everywhere. We get them from human contact. We can pick them up from walking barefooted, or in the food we eat. However, when parasites are out of control and become toxic, we have health problems. This is what happens in the case of the candida fungus. Its presence in the intestine is normal until overgrowth occurs through autointoxication. Then it develops into candidiasis, which is abnormal. Those familiar with the practice of colon cleansing know there is a specific place in the colon known as the region of parasites.

In order to remove the parasites, it is essential to embark upon an effective elimination program. Parasites can be destroyed harmlessly and without dangerous drugs. Ingesting protease enzymes safely accomplishes this. They are nontoxic and natural. Great diligence is required on the part of the client. You must take the capsules with consistency and a real dedication to your cleansing program. When you do, the rewards are enormous.

Cancer: Is It a Deadly Disease?

Cancer cells are formed as a result of physical, chemical, infectious, and genetic mechanisms that interfere with the production of cells. Damaged and abnormal cell production robs the healthy ones of nutrients. Everyone reading this book could have anywhere from one hundred to three hundred thousand cancer-like cells active in the system right now. Does this mean you have cancer? No, but it does mean you may have unbalanced cells that can behave like cancer. If the body is undernourished because it is not digesting and assimilating food correctly, a pattern of chronic fatigue and low blood sugar symptoms will manifest. The by-products of improperly digested food in the system block the small and large intestine, causing constipation. This brings on more imposing problems. Fermentation, putrefaction and rancidity of the food take place. Undigested carbohydrates, such as sugars and starch ferment. Undigested protein putrefies, and undigested fats turn rancid in the intestinal tract. Cancer of any type promotes anaerobic cell activity in putrefied or fermented matter. An anaerobe is a microorganism that can live and grow where there is no free oxygen. A normal cell multiplies where oxygen is present, and a cancer cell multiplies without oxygen. These cells are very much at home in an environment of putrefaction and fermentation. Undigested foods create that environment.

When these conditions prevail, the body sends out signals for help. A group of suitable enzymes reacts and disintegrates the cancer-like cell. Enzymes carry away toxic debris. New cancer-like cells constantly reproduce but if we are healthy and balanced, the enzymes can identify and destroy them. Please understand that if we have a functional immune system coupled with good digestion our intestines will remain cancer-free.

Everyone reading this book could have from one hundred to three hundred thousand cancer-like cells active in his or her system right now.

However, if the balance is upset by poor eating habits, drugs that weaken the immune system, smoking, stress, premature aging, and living life with fear, we will have an adverse reaction. This happens when the balance between the cancer-like cells and our bodies' defense systems collapse. Left alone, the cancer-like cells coat themselves with a glue-like protein fiber. This gives them a thickness of fifteen times that of normal cells. The cancerous cells hide under the glue in an attempt to remain undetected. These cells wander through the body, seeking a place where they can anchor themselves. When they do attach to a body part, they rapidly add more of the gluey fiber. This accelerated growth forms a tumor. Cancer cells are both stubborn and clever. They make disguises for themselves called antigens. An antigen is a protein, toxin, or other substance of high molecular weight to which the body reacts by producing antibodies. When the cancer-like cell is destroyed, many times an antigen replaces it. These cells are so persistent that they will even leave a false trail so they can form new antigens elsewhere. When the antigens become more than our immune system can handle, we are in trouble. When the immune system reaches this depressed state, tissue damage occurs.

The cancer cells grow because there is an absence of enzymes to fight them. In addition, they contain an enzymatic substance that allows them to multiply at a high rate. Further, they can protect themselves by creating a fibrous coating. The action of enzymes strips the fiber coating off the cancer cell. The more of these invaders you have in your body, the more enzymes you need to preserve your health.

I offer you this quotation from Dr. Marrow's book, *What Your Doctor Won't Tell You,* (Harper & Row, 1990). "The fiber around the cancer cell is made of protein. Proteolytic enzymes digest this protein coating which allows macrophages (white blood cells) to attack and destroy the cancer cell."

In the book, *Enzymes: The Fountain of Life* (written by D.A. Lopez, M.D.; R.M. Williams, M.D.,Ph.D.; and M. Miehlke, M.D.; and published by Neville Press) I quote more doctors' findings. "Cancer cells are able to form imitations of the suckers of certain organs and can therefore adhere to the cells of these organs. The enzymes can apparently differentiate between the true suckers and the imitations by the cancer cells, and can then break away the adhesion molecules, making these proteinaceous counterfeits less of a problem."

I grant that even this enzyme book cannot make the final statement on the knowledge of enzymes. Nothing is more stable than change. I conclude by repeating my belief that enzymes represent a highly promising modality of medical treatment. They are the medicines of the future. Their use most surely will increase as we learn more about the immunopathology of diseases through ongoing clinical and scientific studies.

Good News:

- Cancer cells are more sensitive to enzymes than normal cells.
- Enzymes dissolve the fibrous coating on cancer cells, allowing the immune system to work.
- Enzymes can diminish the ability of cancer cells to attach to healthy organs or tissue.

Case History

A woman I know had a health history of cancer from her mother's side of the family. On her father's side there was a long history of cardiovascular disease. Even today, her mother suffers from an enlarged heart and valve problems. At the age of twenty-eight, this lady had a hysterectomy. At that time her doctors told her she had a predisposition to abnormal cells. In her thirties, other doctors informed her she had a lump on the breast that showed abnormal cell activity. They advised her to have radiation or removal. Two years later, after tests and analysis, she was told she had colon cancer. Because of her age and the responsibility she felt toward another family member who needed her attention, she decided against treatment or surgery. Instead, she undertook a personal mission to heal her body by changing her diet, her mind, and the negative way she viewed life. And, she discovered enzymes. The more she learned about prevention the more she was convinced that it was not too late for her. Keep in mind that she made all these decisions and did not start her new health program until after she was afflicted by all those diseases. Within six months after beginning the use of plant enzyme therapy, she went to two different doctors for an examination. Both of them gave her the same answer: she was absolutely clear. She is now in her fifties, experiencing great health and enthusiasm for life. I am proud to report that I am the subject of this case history!

Do I believe in bringing balance to the body so it can heal itself? Absolutely. Do I feel that plant enzymes helped me save my life? Without a doubt. Do I suggest that others take enzymes to bring life to the body? Yes. However, I never pressure them to do or take something they do not believe in themselves. If they feel comfortable with the guidance of their doctors, I will honor that. It is not my place to try to force anyone to follow my program. All individuals have the right to make their own decisions about how they wish to handle their own bodies' health. No one else has that right, and cannot do it for them. It is up to all of us to educate ourselves and make our choices based on knowledge and belief, rather than fear and coercion.

I am the first to say that enzymes are not the only answer to good health. Nevertheless, the one thing we all must do to rebuild and heal is feed and fortify through efficient digestion. Every food we consume has to be digested, assimilated, and utilized as fuel if we want to continually make new, vital cells.

A few years ago, scientists did not know about phytochemicals. Yet today they are the new frontier in cancer-prevention research. The word is derived from the Greek, meaning "plant." *Newsweek* Magazine dedicated five pages to an article titled "Beyond Vitamins" in its April 25, 1994 issue. It stated that raw vegetables and fruit such as tomatoes, apricots, and broccoli contain thousands of phytochemicals. These substances work by waking up enzymes inside cells and detoxifying cancer-causing chemicals.

This article came out just after *The New England Journal of Medicine* delivered a pessimistic message about those vitamins known as antioxidants. The holistic health community's long-awaited verdict in the "great vitamin debate" did not give the vitamin industry the break it wanted. Nonetheless, scientists reported information on a "magic pill," one that would go beyond vitamins. A compound called sulforaphane, derived from broccoli, cauliflower, brussels sprouts, turnips, and kale was discovered that kept laboratory animals free of cancer. Early in the nineties, research with cells in a lab dish led biochemists to suspect that sulforaphane works its magic by boosting the activity of phase two enzymes. These enzymes detoxify carcinogens by hooking them up to molecules and moving them out of the cells where they are rendered harmless.

A phytochemical in strawberries, grapes, and raspberries also neutralizes carcinogens before they can invade DNA. It is called ellagic acid. Even if these early preventatives fail, there is yet another pre-cancer instance where phytochemicals can intervene. In 1993 German researchers announced that they had isolated a chemical in soybeans that prevents the attack of cancer cells in the bloodstream. This chemical is a growth inhibitor of cancer.

Plant enzymes have the same capabilities as these compounds like sulforaphane or the phytochemicals. Their action in the stomach assists the digestion and breakdown of the foods. Ingested on an empty stomach, they can purify the blood. Enzyme action is very much like cellular musical chairs. These compounds race to sites on the cell where cancer-causing hormones, including estrogen, attach themselves. When the music stops, the enzymes keep the hormones from sitting on the cell's surface. The protease enzymes do even a greater job since they can break down the surface or coating of the cancer cell and expose it to the host to immune system.

After the information came out about the research on phytochemicals, several new products appeared on the shelves at health food stores. These products contained freeze-dried vegetables. Claims were made that salads or juices could be ingested in capsule form. In truth, we have known about research and information concerning plant enzymes for over sixty years. When these products are touted as new ones, it really means they are brand new only to the writer.

Cardiovascular

Cardiovascular disease is not limited to the elderly; age has nothing to do with it. Enzyme therapy is used successfully to treat diseases of the arteries and veins. In their book, *Enzymes: The Fountain of Life*, Drs. Lopez, Williams, and Miehlke offer evidence for this. Studies continue to a great extent in Europe, and some in the United States, supporting the use of enzymes. To quote Dr. Inderst in a lecture September 12, 1990, "Enzyme Therapy is a proven treatment and method for disease relating to blood vessels. It diminishes the swelling, activates the system, and stimulates cells such as macrophages (white blood cells) without the long term side effects of drug therapy."

Besides being the number one killer disease in America, heart, or cardiovascular disease, is one of the leading causes of disability. How can this happen? To best understand this complex and critically important part of the body, let us discuss the cardiovascular system.

The primary focus of this system is to move blood throughout the body. It is made up of the heart, blood vessels, and lymph vessels. The circulatory system transports oxygen, nutrients, immune system secretions, hormones, and chemicals necessary for normal function and activity of the organs. It also carries away waste products and carbon dioxide. This miraculous system equalizes body temperature and helps to maintain normal water and electrolyte balance (salt and potassium).

An adult has an average of 6 quarts of blood in his body. The circulatory system carries this entire quantity through one complete circuit of the body every minute. In the course of twenty-four hours, 7200 quarts of blood pass through the heart. Very small vessels known as capillaries are so tiny the blood cells must enter them single file. The rate of blood flow through the vessels depends on several factors. These are: force of the heartbeat, rate of the heartbeat, control of the quantity of blood that enters the heart chambers, and finally the capillaries. The blood then returns to the heart by flowing through the veins, and the process begins again. The circulation of the blood is extremely important. As it passes through the arteries and veins in the heart, neck, head, brain and digestive organs it delivers and picks up carbohydrates, proteins, fats, and chemicals. When blood enters and exits the liver it brings glucose and glycogen and removes toxins. In the kidneys, it cleanses. In the lymphatic glands, it delivers nutrients, and in organs like the spleen, tonsils, and thymus, it supports the immune system.

The organs and systems of the body vary greatly in the quantity of blood they require at different times. For instance, the brain needs a great deal of blood all the time. Muscles demand blood sporadically. Many things depend on physical exertion, which increases the blood flow to the muscles. Hot weather increases the flow of blood to cool the body. Whenever we eat a meal, extra blood is required by the stomach to help digest and absorb the food. The nervous system controls all this activity. Without this vital control, none of the various needs for blood could be maintained. If the nervous system was not doing its job, the quantity of blood in the brain would change if you shifted the position of your body even slightly. There are two nerves that control blood supply. One is located in the heart; the other is in the neck. These two nerves are perfectly synchronized. They register any changes in the blood pressure and cause the nervous system to change the rate of heartbeat and the size of the blood vessels to always maintain correct blood pressure. Perhaps now you understand the importance of the circulatory system and how our very life depends on its precision work. It is yet another of the marvels and miracles of our bodies.

Lipase is the major enzyme used for the break down of fats and cholesterol, and lipid regulation. Recently, a cardiologist whom I know admitted that if the public understood the capabilities of lipase there would no longer be a need for him to continue his medical practice. I thought that was quite a statement! Obviously this is overstated since hereditary heart diseases would always create a need for cardiologists. Lipase is used with other plant enzymes to assist with proper digestion. This feeds and cleanses the life-giving blood in the body.

If the public understood the capabilities of lipase there would no longer be a need for cardiologists.

Enzymatic action runs the life force in our bodies—every thought, every breath. Our entire being is enzyme-dependent. Do I believe enzymes are important? What do you think? Can they assist in the breakdown of food, delivery of nutrients, and fortification of the blood? If used correctly, are they the very life source for our kidneys, liver, brain, and lungs? I answer an absolute yes to all these questions!

My observation of those with heart problems suggests that elimination of sugars and simple carbohydrates from the diet is an excellent starting point.

Frequent urination is a difficulty they face. This is because their doctors have prescribed diuretics to help control hypertension and a disturbed heart rhythm. However, when the client gets rid of excess sugar consumption, the blood sugar level begins to stabilize. Then the body can start to eliminate toxins naturally. The medications being taken are reduced to a lower dosage. Heart disease is an area of great interest to me because my mother and my aunt suffer from it. When I'm with my mother I have noticed that control of her diet has a direct effect on her heart rhythm and how often she urinates.

Blood vessels are just as biologically active as the liver, kidneys, or any other vital organ. The blood vessels depend on enzymes. When the blood vessels are clogged with fats the effectiveness of the enzymes is reduced. The blood vessels can become constricted. How can someone who is interested in prevention deal with this? Keep your enzymes working at all times. Everything that happens in the body does so because of these miraculous substances. As we become more stressed or ill, our immune systems suffer and the power of the enzymes is seriously undermined as well. It makes so much sense to take plant enzyme supplements in order to feed and fortify the body. This takes pressure off the digestive system, allowing it to function more efficiently. Plant enzymes keep our blood healthy, and keep our systems in perfect balance.

Arthritis

Forty million Americans—more than all heart disease, cancer, and diabetes cases combined—are afflicted by arthritis. Arthritis is a family of diseases. Treatment costs the nation one hundred fifty billion dollars every year. Incidence of it is expected to skyrocket as baby boomers begin to turn fifty. By the year 2020, sixty million people will have some form of arthritis, states Dr. Charles Helmick of the Centers for Disease Control and Prevention.

Women are especially susceptible. Eighteen percent of women have arthritis compared to only 12 percent of men. Nearly five million women are forced to limit their activities as a result of this disease. Medical terminology uses the suffix "-itis" to describe disorders involving inflammation. A partial list of these includes arthritis, pancreatitis, dermatitis, and prostatitis.

Rheumatoid arthritis is systemic, meaning it is present throughout the body and involves the connective tissue. Degenerative lesions may be found in the collagen (a fibrous protein in connective tissue, bone and cartilage) of the lungs, heart, blood vessels, and pleura. Those who suffer from the rheumatoid variety often appear to be undernourished and chronically ill because of the inflammation of blood-forming organs. Most common is the arthritis that attacks the joints.

Osteoarthritis is located in the musculoskeletal system (muscle and skeleton). Arthritis is one of the most difficult conditions to treat. The drugs used for treatment stop the body from making prostaglandins. While prostaglandins are responsible for the swelling of the joints, they also protect the stomach from acids and other digestive juices necessary to dissolve food. As a result, about 15 percent of all arthritic people have ulcers. The goal for medical technology is to develop a drug that reduces swelling without disturbing the stomach. Many of the other treatments have harmful side effects that lead to other diseases. No matter what the treatment or side effects are for arthritis, the one certainty is it interferes with quality of life.

The consistency of the blood changes and more white blood cells are created as a result of inflammation. This is how the body naturally reacts and attempts to protect itself. A positive response to inflammation depends on adequate enzymes. If the inflammation is constant or the onset of it is especially severe, an increased abundance of internal support enzymes is required. However, the quantities needed are too great for the body to produce without assistance. The body is unable to radically change its rate of enzyme production, just as it is incapable of drastically slowing it. The significance of a constant, steady supply of enzymes cannot be overemphasized.

When the body's immune (defense) system breaks down, it is because of an internal enzyme shortage. This begins a process called autoimmunity. The body begins to regard its own tissue or cell structure as a foreign invader. Our immune system begins to fight against itself. It attacks its own connective tissue, joints, skin, and blood cells. While this battle rages, the original disorder is still taking its toll. The outcome is chronic illness.

In autoimmune disorders, I see some common denominators. Most of those who suffer have problems digesting fats and protein. They tend to be either Body Type Two or Three. They prefer heavily spiced, strong flavored foods, crave sugar and were raised on meat. Does this mean they will never eat meat or spicy food again? No, it means that these are the cravings of most arthritic people I have counseled. Protease and lipase are the enzymes they need the most.

The case histories are numerous. They are all very much the same. This is why I approach all the autoimmune disorders with the same enzyme regimen. First, I address the primary problem of poor digestion with a multiple-enzyme supplement taken with every meal. High dosages of protease are taken between meals for the antioxidant properties. Most of these clients are constipated. I suggest the use of a bacterium-fortified enzyme with herb formulation to assist with frequent, healthy bowel movements. Without exception, all systems are very toxic. With the aid of plant enzymes we feed, fortify and detoxify.

Immune Boosters: "The Enzyme Connection"

In thousands of my clients' case histories, enzyme supplements produced stunning turn arounds. These amazed not only themselves but their families, friends, and even their doctors. I am not alone among health-care professionals who have experienced this. Many of them have shared the same information in interviews, books or by word-of-mouth. It seems as if enzymes play two different roles in rebuilding the immune system. First, they act as an agent that counteracts the causes of the initial damage. Then, they enhance the immune system itself.

This is how it works. Digestive enzymes attack the offenders themselves by reducing inflammation and neutralizing many of the toxins that injure the mucosal wall. This reduces leakage of partially digested food remnants, mostly chunks of allergenic protein, through the intestinal walls and into the bloodstream. Our powerful protein-digesting enzyme, protease, comes to the rescue. When taken on an empty stomach, it escorts the protein fragments from the small intestine into the colon where they are destroyed through normal digestive procedures. This same enzyme, when no longer needed in the intestine, moves directly into the blood. There, it continues to remove toxins and other unwanted protein matter.

Plant enzyme supplements help strengthen the immune system by increasing the number and intensity of the fighter cells. These warrior cells are macrophages, monocytes (large, mononuclear, nongranular white blood cells with round or kidney-shaped nuclei) and T-cells (lymphocytes involved in rejecting foreign tissue, regulating cellular immunity, and controlling the production of antibodies). When the enzymes reach these immune-system cells, they increase the fluidity and permeability of the cell membranes. This makes it much easier for nutrients to enter the cells . . . and much easier for toxins to leave. This creates improved cellular metabolism throughout the body and an overall boost in energy level. After the vicious cycle of malnutrition, poor digestion, and cellular starvation is finally broken, your immune system is remarkably stronger.

It has been my fortune to see the changes that enzymes make in the human blood. Working with medical doctors, I have the honor to be a nutritional consultant to some of them. This gives me the opportunity to study the experiences and case histories of many clients and patients. I feel great humility when I see an enormous transformation occur in someone who has been critically ill. One gentleman, in particular, comes to mind.

Case History

A successful surgeon called my office on a Thursday afternoon. I clearly remember the day of the week because it plays a role in this story. The surgeon asked if he could bring one of his patients in to see me. He wanted to observe the Live Blood Analysis. This is a procedure I teach to preventative practitioners that allows them to evaluate live blood samples. It is used as a teaching tool, rather than a diagnostic one. A drop of blood from the finger is placed on a microscope slide. This technique is called dark-field microscopy. The observer can detect motion in the cells, noting balances and imbalances.

The patient in question had just completed his final chemo treatment. With a dangerously low white blood count, his doctors were afraid to give him any more. The injections he received to elevate the white count did not work. His WBC (white blood count) was five hundred eighty. Normally the leukocytes should read about eight thousand WBC. His was so far below normal that his physicians were concerned he might have a compromised immune system.

I knew, of course, that the enzymes would elevate the WBC. However, the surgeon was desperate. He wanted to see a detectable difference almost immediately. The doctor intended to do another blood count on Monday morning. It was late Thursday afternoon. When someone is relatively healthy, at least not suffering from terminal illness, we can see an increase in both the WBC and RBC within ten minutes of ingesting a protease simulation. In this case, we were able to see some new white blood cells and T-cells in a matter of a few minutes. Both the doctor and I were elated. I explained to his patient that I had no idea until that moment that we could see results so quickly after chemo. As they were leaving, I gave the man nutrition information along with all the standard enzyme formulations. By Tuesday morning, a very happy doctor had phoned me to inform me his patient's white blood count was back up to seventy-eight hundred by the previous afternoon. The last time I saw this man, he was doing very well, his cancer is in remission. His hair was growing back in and he told me he felt wonderful.

Do You Have Healthy Blood?

Do you have healthy and vigorous blood? Or is it tired and weak? The answers to these questions could very well mean the difference between a healthy, energetic lifestyle and one of constant weariness and fatigue.

We cannot live without that extraordinary and complex protein fluid coursing through our bodies. A miracle of modern science is our ability to use one another's blood as long as it's the same type. The blood's liquid component is called plasma. It is the primary ingredient of this life-sustaining, homogenous mixture. The plasma functions as the carrier of food, oxygen, vitamins and minerals to cells, in exchange for waste matter. The solid matter in blood consists almost entirely of coin-shaped red cells. Remarkably elastic, they are so flexible they can extend, fold up, and squeeze through a capillary barely half their own diameter. Red cells are naturally manufactured in our bone marrow. Their principal ingredient is hemoglobin, a protein molecule programmed to combine with oxygen.

Blood contains other miraculous substances as well. One is a thin, plate-shaped cell, called a platelet, that is essential for clotting. If the skin is cut, for instance, and a blood leak occurs, the platelet releases a chemical that crystallizes one of the plasma's proteins. Called fibrinogen, it creates a net or a web of red cells, all hooked together into a solid plug. The larger white blood cells, considerably less numerous than the red ones, act as the body's defense team. When an invasion of bacteria, fungal forms, or parasites occurs, white blood cells multiply as they engulf and destroy these enemies.

Our blood cannot survive or function without oxygen. The manner in which blood is oxygenated is remarkable. The average adult male has a capillary system that would stretch out, end to end, to about 60,000 miles. The lungs are fraught with these minute vessels. As we take a breath, the oxygen we inhale enters the blood through the very thin walls of these capillaries. The blood carries this oxygen to all the body's tissues. Capillaries are like garden hoses inserted into the river of an artery or vein. Each one is less than 1/1,000 of an inch in diameter. More than three million of them thread through each square inch in the cross section of a muscle. Each capillary carries its dribble of blood swiftly. The heart patiently pumps the body's 6 quarts of blood through all its tissues again and again. The average pace of a complete round trip is about one minute.

The lymph system is a companion to the circulatory, or blood system. Its primary purpose is to isolate and eliminate dangerous infections. The lymphatic system circulates more or less parallel to the blood. The lymph glands, or nodes, are made up of valves and filters.

Blood circulation is continuous but not constant in an active body. Muscular, electrical, and chemical controls unceasingly regulate and change the pattern of flow. It acts like a sprinkler system in your front yard. Different groups of sprinkler heads open and close, flooding first one area and then another. Blood can flow like a big river, causing obstacles too big or heavy to be flushed away, sometimes even reversing the current in certain channels. The entire operation is controlled by special centers in the brain. These receive information from sensory monitoring devices located at strategic points. They send signals to the heart and to thousands of arterial, venal, and capillary control stations.

Human blood, in a nutshell, performs all these life-giving functions for our bodies:

- Circulates oxygen throughout the body's network of arteries, veins, and capillaries
- Delivers nutrients to the tissues and organs
- Carries minerals, hormones, vitamins, and antibodies
- Removes waste products

Many substances vital to health are recycled through the blood. Blood helps maintain equilibrium (homeostasis) of the internal environment. It bathes the tissues with oxygen. It collects waste products. Its major regulatory duties involve the nutrition of cells, protecting against foreign invaders, and keeping the body temperature constant. The blood also facilitates the body's adaptability to changes in external conditions. These include climatic changes, stressful physical activity, new eating habits, and resistance to injury and infectious organisms. Blood is analyzed for information leading to nutrient deficiencies and disease.

Red blood cells circulate in the bloodstream for about one hundred twenty days. After that, they are trapped and broken down in the spleen, an organ responsible for storing and filtering blood. The spleen and liver can function as backup sites for red blood cell reserves in an emergency. They act as salvage yards for iron reclaimed from dead red blood cells. The total numbers of red blood cells in the body vary with age, altitude, activity, and temperature. An average, healthy individual has about thirty-five trillion of them.

White blood cells or leukocytes are the principal component of the body's immune system. Acting as scavengers, they assist with the repair of the body. They are generated in the bone marrow or lymph nodes. These nodes serve as defense outposts against germs attacking the body. The white blood cells travel from site to site through the arteries, veins and capillaries. White cells can also leave the bloodstream and filter into the lymphatic tissues if necessary to fight infection. Although these cells are produced in large quantities, they die within a few days. Vastly outnumbered by the red blood cells, the average, healthy human has only about seventy-five billion leukocytes in the blood.

Personal physicians order most of the blood tests performed on human beings. The blood taken from the inner bend of your elbow at your doctor's office is almost always sent to medical laboratories for testing. However, there are ways for the layperson to observe his own live blood specimen. One method is called dark-field microscopy. I remind you that this is not used for diagnostic purposes. Rather, it is a nutritional teaching tool. I use it in our research programs. Through the powerful lens of the microscope, we can observe white blood cells to see if they are doing their job efficiently. This can be an excellent indicator of the performance of the auto-immune system. Imbalances detected by the dark-field microscope are based on nutritional information about the client, which shows up in blood activity. Someone who has problems digesting fats could have a buildup of platelets adhering to the walls and rough surfaces of a damaged vessel. This condition indicates high cholesterol and headaches. Poorly digested protein will show up in the plasma and will cause the red blood cells to stick to each other. When the liver is toxic from alcohol consumption, prescription drugs, or lack of digestion, it shows up as fibrin material in the plasma.

Taking mega-doses of vitamins and minerals can create imbalances. Our systems are overwhelmed when we take more than we need.

When digestion is actually in progress, blood fats are easily recognized by the dark-field method. Bowel toxicity appears as bright, shiny crystals. Often, these are so toxic they will show up in color on a black and white screen. Some deficiencies look like oval or misshapen red blood cells. In the case of a nutrient deficiency, the red blood cells cannot maintain form and simply fade from view. White blood cells are easily distinguishable and give information about the immune system. One of the most commonly occurring problems is a variation in the white blood cells, or leukocytosis. These cells are present in the blood because the body is in a state of alert against the havoc caused by poorly digested food. The body sees the undigested particles as invaders and responds by calling out the white blood cells.

Dark-field microscopy was instrumental in proving to me the importance of plant enzyme therapy. I tend to try new products on myself as a living laboratory. After ingesting the product, if I observe a change for the good, I am interested in further use of it. Unfortunately, many formulations do not show a positive change. In my own experience I have discovered that too often those taking mega-doses of vitamins and minerals can create further imbalances. This can be just as much of an upset to the body's nutrients as it is when poor nutrition occurs. A practioner can measure these imbalances by using blood, urine, and saliva. Our system is overwhelmed when we take more nutritionals than we need. Ingesting enzymes are a different matter altogether. They are the catalyst to make things happen in the system, they are not stored nor can they cause an imbalance.

The enzyme supplement regimen for keeping the blood healthy is simple to follow. Please remember that plant enzymes are an approved food item by the FDA. As of this writing, the measurements have yet to be standardized. Most companies have the integrity to use the correct amounts when producing the capsules. These are the amounts I suggest.

Digestive formulations per capsule:

- **Amylase**: 5,500–12,000 units of activity
- **Lipase**: Minimum 145–500 LU for fat digestion
- **Protease**: Minimum 30,000 HUT
- **Cellulase**: Approximately 600 CU (can vary). The body does not make cellulase. If you have problems with fiber and grains, you want at least this dosage.
- **Lactase**: Minimum 300 LacU for dairy product intolerance.

This formulation should also include some maltase and sucrase. The quantities will vary. They are not as important as protease, amylase, lipase, and cellulase.

Immune support formulations per capsule:

Protease: Minimum 375,000–420,000 enzyme activity. Take between meals, 3–5 times daily.

Any dosage above 40,000 Protease should always be coupled with 8 ounces of water. If you need an extremely high dosage of protease you will need to take many capsules at once to fulfill the requirement. Only a health-care professional will carry a product line that comes in such enormous amounts. These are therapy dosages. Protease is the strongest scavenger antioxidant on the market. By comparison, the digestion dosage is approximately 30,000 Protease HUT or more. It is so important for a formulation to contain pharmaceutical grade pure plant enzymes. If it does not, there is never any guarantee that nutrients will be delivered to the cells, the building blocks of our body. If you do not thoroughly digest with the aid of these enzymes, you will be like most Americans. We all have very expensive urine and fecal matter.

Chapter 4

Enhance Your Mental Capacity

The brain is the exchange center of the nervous system. This is where sensations generate ideas, and ideas are expressed by action. Evidence of its importance to the body comes from the fact that the brain accounts for only 2 percent of the body's weight; yet its operation uses 20 percent of the oxygen and blood. It's a mass of soft, spongy, pinkish-gray nerve tissue. The weight of the brain in an average adult is about three pounds. The neurons, or nerve cell bodies, fit together as precisely and compactly as a puzzle. They are so intimately coordinated as to allow the size and shape of the brain to be used to maximum efficiency.

Neither the weight nor the size of the brain is a reliable clue to the intelligence of the individual. The size of the cells does not matter, either. Rather, the number of brain cells is exceedingly important. The human brain, and that of higher animals, is not fully formed at birth. Like life itself, the brain is a growing, flowing thing. It first becomes discernible in the human embryo about two weeks after conception. A thin, gray film of cortex begins to spread across its upper surface. It crinkles as growth continues and it nestles into the cerebrum. At birth, the human baby is about one-third head and literally top-heavy with brains. The infant brain is still only an overgrown seed, weighing in at only 25 percent of its adult weight. Just as muscles grow with exercise, so does the brain as the child gets older and uses the brain's countless parts and pathways. The structure of the brain is formed by the mental experiences of the growing child. This formation is the most rapid in the preschool years. However, it continues to a lesser degree throughout life. About 50 percent of the brain's capacity is genetically inherited, while the other 50 percent is culturally shaped (by teaching and example). Day by day, the living brain models itself biologically as a physical organ. Language, images, and ideas continually course through its convoluted passages. The brain is protected from harmful substances in the bloodstream by the blood-brain barrier. It keeps most toxins away from the brain, or at least delays their entry for several hours, sometimes days. If the toxins have already penetrated other parts of the body, they are blocked from the brain.

The hypothalamus, an organ the size of a lump of sugar, is located at the base of the cerebrum (the upper, main part of the brain, consisting of the left and right hemispheres). Although it is very small, the hypothalamus controls some very vital functions, including the ebb and flow of the body's fluids, the regulation of fat and carbohydrate metabolism, blood sugar levels, and body temperature. It directs the rhythmic cycles, such as activity and rest, appetite and digestion, sexual desire, and the menstrual and reproductive functions. In addition, the hypothalamus is the body's "emotional" brain. This miraculous little organ is the coordinating center for the nervous system. It controls our sleep, wakefulness, alertness, and reactions to pain and pleasure. Without this gland, I would not be writing this book—everything I discuss here depends on the hypothalamus!

Now, I will take a look at the connection between the brain and the body, and how digestive plant enzymes support them. Food is defined as anything ingested into the body that serves to nourish or build tissues and cells. Doctors William H. Philpoot, M.D.; and Dwight K. Kalita, Ph.D; in their book titled *Brain Allergies* (published by Keats Publishing Inc., copyright 1980), state "A person cannot be nutritionally deficient, toxic, infected, or addicted to any food or chemical without suffering the consequences of progression of the disease process into a chronic degenerative illness of some type."

We have been conditioned by the media and others, for years, to eat a candy bar or drink a cola beverage for that boost around three o'clock in the afternoon.

It is vital to feed and fortify the hypothalamus in order to have a healthy, balanced body. Many uninformed people will naturally assume this means eating sugar to energize the system. We have been conditioned by the media and other sources for years, all of them telling us to eat a candy bar or drink a cola beverage for that boost around three o'clock in the afternoon. However, brain food must be highly nutritional and must contain high levels of oxygen. If we have digestive and assimilation difficulties, the brain will suffer. When food remnants in the bloodstream reach the brain, the barrier regards them as invaders. The hypothalamus operates our endocrine and nervous system, our two major control centers. It stands to reason that we feed the brain only the healthiest of foods.

Case History

One day a woman brought her husband to my office, a gentleman in his sixties. His son, a medical doctor, had tested him for Alzheimer's. Hoping to find a natural way to treat him, they requested diet and enzyme counseling from me. I suggested a diet suitable for his body type and enzyme formulations to feed and fortify his system. Keep in mind there is a time to feed and fortify and a time to detoxify. However, the beauty of enzymes is that these formulations allow us to do both at the same time and in harmony. I put him on a vigorous enzyme and food plan, and they departed for their home in another state. Three weeks later I received a phone call, followed by a letter. The elated wife told me that before beginning his enzyme program, her husband could not remember their children's names, and could no longer read or watch television. After only a few weeks on enzymes, her husband wrote notes to her and was watching TV again. Within nine weeks he was almost completely restored to his former self. Is this a miracle? Yes . . . it is the miracle of enzymes delivering nutrients to a depleted system that desperately needed them!

Fibromyalgia

FM is a "chronic invisible illness." It is not just a form of muscular rheumatism. It is actually a type of neurotransmitter dysfunction. One symptom is poor concentration. Exactly what is the nervous system and how does it affect our concentration? It is composed of the brain, the spinal cord, and all the nerves. Collectively, these work in harmony with one another and serve as the communication and coordination systems of the body. The nervous system transports information to the brain, and relays instructions from it. There are two subsystems that together perform more complicated functions. The first is the autonomic nervous system. The caretaker of the body, it operates without conscious control, taking care of involuntary actions like breathing or the heartbeat. The voluntary nervous system includes both motor and sensory nerves. Among many other tasks, it controls our voluntary muscle movement and carries information to the brain. The command centers of both these systems lie in the hypothalamus. They are in perpetual activity, fine tuning the body's processes to all the internal and external demands made upon it. Research is showing that people with FM have defects in the neuro-regulatory system, especially transmitters. There seems to be a genetic predisposition.

How does any imbalance of the nervous system affect our digestive system? You have heard many times that it is not a good idea to eat when you are upset, or if you have suffered severe trauma. When the body is stressed, many reactions are set into motion. Elevated levels of epinephrine are released into the blood to enhance muscle action. Glucose is immediately dispatched by the liver. This is a source of instant energy for all the muscles. Heart rate and breathing accelerate, but digestive activity slows. Blood vessels constrict to cause sweating, so the body will stay cool while under stress. Now, the body is ready for some extraordinary actions.

While all this is taking place, the parasympathetic nervous system prevents these processes from speeding up to extreme rates. The parasympathetic, part of the autonomic system, originates in the midbrain. It monitors constriction of the eye pupils, slows the heartbeat, and stimulates certain digestive glands. The parasympathetic acts as a damper. If the trauma is not too serious, it allows the body to return to normal. It does not work as fast as the sympathetic system (also part of the autonomic system, which works in opposition to the parasympathetic, *i.e.*, dilates the pupils, etc.). Messages from the brain, often in response to information conveyed by the sensory nerves, are delivered to the muscles by motor nerves. One motor nerve with its branching tendrils may control thousands of the muscle fibers.

All the parts of the nervous system are constantly interacting with one another. They are so well coordinated you can simultaneously think, feel and act on many levels without confusion. We suffer neurological disorders if the organs of the nervous system are inflamed or affected by disease or injury. These manifest as paralysis, seizures, or sensor malfunctions. When we say we are nervous, it simply means we are in a state of excitability with great mental and physical unrest.

Our nervous system is fed by the foods we eat and the enzymes carrying the nutrients. The enzyme action is what helps the nerves communicate with each other. Inflammation creates problems for the nervous system, the symptoms for which can be tenderness, pain, or paralysis of a limb. Toxic substances poison our nerves. Alcohol and any vitamin deficiency are the worst offenders. Diabetes poses additional difficulties for those with nervous disorders. Any type of allergy, as well as some viral or bacterial infections have an impact on the nervous system.

Rest, therapeutic enzymes, and a nutritious diet are the best treatments for these conditions. Extra vitamins from the B group are essential. However, they cannot be delivered without enzymes. Very few of the vitamin tablets on the market are chemical-free. Even enzymes cannot deliver artificial compounds.

Foods called super foods, those highly nourishing ones that occur naturally, are those whose nutrients will actually be delivered. Chemical vitamins usually create more toxicity rather than provide any nutritional benefits. So if you take vitamins long term, make sure they are natural complex vitamins, not synthetics. For example, B6, B12, etc., should say Whole B Complex, not cutout with a chemical and measured as in B6–50mg.

Anxiety

Anxiety is a feeling of uneasiness, apprehension, or dread. In many instances anxiety is a perfectly rational response. We may feel anxious about passing an exam at school, how well we will do at our new job, or moving to a new city. Realistic anxiety means we have pronounced concerns about world issues like war, the effects of toxic air, or social and economic conditions that affect our lifestyles. Our easy access to the mass media can serve to intensify normal anxieties. A certain amount of unrealistic and irrational anxiety is an accepted part of our daily lives.

Our awareness of both past and future can result in increased levels of anxiety during certain times of our lives, especially adolescence and middle age. People who spend much of their time alone are likely to suffer more anxiety than those who live and work with others. Most of us find healthy ways to work with our anxiety. We make new friends, take up new hobbies, become good listeners, and volunteer in our communities as a means of helping others. We can relieve our anxiety through exercise and physical activity. Games and sports, participating in group activities, or taking a good walk often dissipates an anxious mood. Real anxiety is created by something specific and identifiable, such as alcohol. When anxiety is chronic and not traceable to a specific cause, we call it neurosis. This is an especially difficult one, because it always interferes with normal activity. Often the sufferer is subject to extreme fear. It becomes pathological when the individual can no longer control his emotions. It may be severe enough to manifest as organic pain or true physical illness. The best way to fight anxiety is to find its cause. Identify your own specific body type and investigate your food allergies and choices. This will greatly help you to alleviate the problem.

Enervation: The Cause of Anxiety and Nervousness

Enervation is physically induced anxiety. We can enervate ourselves in a variety of ways. Overeating, too much alcohol, sugar, cooked food, salt, caffeine, tobacco, drugs, and impure water consumption, as well as over-work, worry, tension, depression, and lack of rest all serve to enervate us. When these habits are dropped, headaches and general letdown usually occur. As the body discards toxins, it is normal to experience depression, a decrease in energy and feeling generally unwell. You have heard the expressions, "You need to feel bad to feel good" or "You will get a lot sicker before you get better." Several times when treating clients, I have found that these beliefs have kept them from giving me an accurate picture of their health. They assume they should feel unwell, and it is not important enough to let me know about it. After realizing what was happening, I could match the enzymes with their needs. This fortified their bodies, creating a greater sense of well-being and increased stamina. I believe a great deal of the disease we see in our society today is caused by enervation. Enervation resides in almost every home in the country.

Years ago when I first began my practice in nutrition the first two weeks of association with a new client required great vigilance on my part. During those times, the client was in desperate need for motivation. Many of them were, at first, under the false impression that they did better on junk foods. It was my responsibility to get them through the first stages of withdrawal. I heard complaints of weakness, of feeling a lack in the necessary strength to continue. Often the weakness and the results of the dietary changes were blamed on me! I had to do some fancy footwork to convince them of the need for "house cleaning." Sometimes it was very difficult to accept the fact that it was normal to feel ill. The more toxic the body, the worse they felt.

Now, I realize that when a client is experiencing cleansing and still feels ill and fatigued, it is because I did not match the correct amount of plant enzymes to their needs. The body will move through a catabolic state. It rids itself of old obstructive material that has been stored in the tissue and joints. However, when we ingest the correct amount of foods high in nutritional value, renewed strength and rejuvenation occur. I like to see each client two weeks after the initial visit. I know the client will be in a recuperative period and that the enzymes will have done a good part of their work. Most of the clients' education takes place at the second visit. Frequently, they have more questions about diet. This establishes a working knowledge between us of where we are in their wellness cycle. One month later, I see them a third time. Where they are now in an anabolic (constructive metabolism—the process by which food is changed into living tissue) stage of healthy living. From time to time, as you continue to build good health, you may experience common cold or other symptoms. If this occurs, simply return to the routine initially prescribed for the enzymes. This type of recurrence is

There are no incurable diseases; there are only incurable people.

usually due to not taking the correct quantity of protease while detoxifying. I am very committed to the principle that the more clients are educated and encouraged, the more likely they will take responsibility for their own progress. I have an open-door policy. I always want my clients to feel free to ask questions about their healing and their state of balance.

As a teacher in the field of building good health, I have discovered I must have patience, thoroughness, and kindness. These are easily recognized traits. If they are not found, the prospective client will go elsewhere. This does not mean the he can get away without assuming any responsibility for himself. When I think of someone about to undertake a natural health regimen, it brings two old adages to mind: *"You can lead a horse to water but you can't make him drink,"* and *"There are no incurable diseases; there are only incurable people."* Prospective clients come to me after they have doctored themselves for years by taking various medications. Yet, they are still miserable and sick and never really feel good. When they hear about the enzyme program, they expect to feel like new overnight. That will not happen. Any time you decide you are going to start a new, healthy lifestyle, you must be very committed. If the program is going to work, it will not be a fad or trendy diet. Rather, it will be tailored to your needs and your body type. You cannot change your body type, who you are, what your family is like, your strengths and weaknesses by simply changing your mind. Some things we bring into our lives are hereditary. As we get closer to a balanced state, it may be more difficult to recognize our body type. I ask each client to take the enzymes for at least six to nine weeks from the initial visit. As they progress, I will start them on a maintenance program.

Education begins when enervation is detected. Rebuilding a consciousness of wellness is a good starting point. People who suffer from anxiety often forget to breathe! Since oxygenating the blood is critical to good health, it is no wonder so many of us feel ill and fatigued all the time. The purpose of breathing is to supply oxygen to the body and to aid in cleansing the blood stream. Breath is life. It has been said that we learn to breathe a few seconds after we are born, and when we forget to breathe, we die. Abnormal cells multiply in the absence of oxygen. The significance of proper breathing and fresh air cannot be over-emphasized. Some have a problem breathing because of an acid/alkaline imbalance caused by poor food choices for their own unique body types.

What happens when we breathe? The chest rises and falls while the diaphragm expands and contracts. This naturally creates a vacuum into which air rushes. Finally, the air is forced from the lungs. Blood then enters the lungs to pick up the oxygen and carry it to the cells. At the same time, it takes on carbon dioxide that has been expelled by the cells. The entire process is automatically controlled. This respiratory center is located in the medulla oblongata at the base of the brain where the top of the spinal cord widens. Carbon dioxide acts as a stimulant of the respiratory center. The more carbon dioxide present in the blood, the more the respiratory center is stimulated and the faster we breathe. Conversely, the more oxygen in the blood, the slower we breathe because oxygen inhibits the medulla. Our breath rate and volume are constantly adjusted to the body's needs at every single moment. If the air around us is clean, it gives us energy. I encourage everyone to spend time outdoors, breathing fresh air as much as possible. We are living beings and breath is life.

To go beyond enervation we must change the way we eat and change our attitude.

I closely observe people as they relate their stories of illness to me. Many times they rush through these tales, never stopping to breathe. It is sad to me how our society has become so anxiety-ridden. May I say again, the way to make a change is to make a change! Do the things you find pleasurable. To go beyond enervation, we must change the way we eat *and* change our attitude.

Attitude

I will never stop saying how crucial it is for you to develop a good, positive attitude. This is of extreme importance in the practice of good health. Remember, there are no incurable diseases; there are only incurable people. I know people who would rather sit around all day and talk about their illnesses than do anything about them. Why? Their thinking is misguided. They are self-absorbed and negative as a result of self-poisoning and allowing themselves to be brainwashed. These are the ones who think ill health is normal. We discuss our health in everyday conversation as much as we talk about the weather. Ever since the days of Freud, we have been taught to cure our problems by dwelling on them. The efforts of psychiatrists and psychologists today help to perpetuate this, leaving us with uncured mental health maladies and a prevailing order of failure and chaos. There is an above-average rate of suicide among these mental health practitioners. Dr. Abraham Maslow, a brilliant psychologist, has changed all that. He said, "What we need is a new direction. We ought to stop dwelling on the sick and the deranged if we want to restore people to a perfect

We ought not to treat illness but to direct the person into the life patterns of the mentally healthy.

mental health. We ought to study health; not illness. We ought not to treat illness but to direct the person into the life patterns of the mentally healthy." Those who are using Dr. Maslow's philosophy are revolutionizing the mental health field with tremendous results. I have found it is far more effective to talk about healthy knuckles, rather than the arthritis afflicting a client's hands.

Those who are truly successful in building optimum health are those who never look back, once they have established healthy new life patterns. They focus on constant rejuvenation of the body's systems. The food we eat is our body's fuel. These foods nourish and fortify our nervous and endocrine systems with healthy nutrients. I ask you to remember at all times, we cannot take synthetic, chemical vitamins and expect them to be delivered to our cells. The body is not designed to assimilate chemicals. It will attempt to digest and deliver them, but in the process it uses up precious metabolic enzymes needed to rebuild nerves and tissues. Plant enzymes will only deliver pure, natural nutrients. Common sense tells us these nutrients should come from one source: pure, natural foods. If you want to nourish your nerves and your immune and endocrine systems, you must ingest a certain amount of protein, complex carbohydrates and some fats. Our nerve fibers are made of protein and fat.

When I treat a client, I am usually working in cooperation with a medical doctor. They have requested my services for providing good nutrition for their patients.

Case History

One instance of this was when a highly intelligent gentleman came to my office. A retiree, he was using his time and knowledge as a volunteer in the community. He was in his seventies. When he entered the room, he wore gloves and moved cautiously when he sat down. I asked him how I could help him. He told me he had a myelin sheath disorder. The myelin sheath surrounds the nerve fibers, and is made of lipid materials, fat, fatty acids, and protein. When the sheath is disrupted, the raw nerve is exposed. The myelin sheath is believed to influence the rate at which nerve impulses are conducted. It can be very painful if these fibers are exposed. Imagine how it would feel to be in constant pain, as this man was. He had to be extremely careful about everything he touched and everything that touched him. This illness was dragging him down. He had tried countless medications and seen many different doctors. In desperation, his doctors suggested he look at nutritional alternatives. What took place surprised both the physicians and me. When I advised him that it only made sense to give him the enzymes his body required, he was amazed. The makeup of the myelin sheath is fats, fatty acids, and protein. He craved all the foods that were rich in these things. In addition, he always had his afternoon Snickers Bar or some kind of candy. This had become his sole source of pleasure. He was not sleeping well, which was perfectly understandable. How could anyone sleep who was in constant pain?

As I began my work with him, I tried to be very gentle. He was the bravest of men and told me to do whatever was necessary. We proceeded as if he was just another client walking through my door. I got to know him, and greatly admired his courage. The most exciting part of this story is that we were able to successfully change his diet. This was a challenge. He was set in his ways, unwilling to change too much. He told me his diet had always worked before, and that was why he was skeptical. I compensated for his resistance by giving him larger amounts of the plant enzymes.

His enzyme therapy was very intense. He took enzymes with meals, enzymes between meals, enzymes first thing in the morning, and last thing before going to bed. I gave him enzyme combinations with herbs that fed and fortified the nerves, his major organs, and his endocrine system. It was remarkable to see how much better he felt when he came in for his appointment twice a month. He was sleeping peacefully and throughout the night. His terrible constipation problem had disappeared. He no longer had such intense cravings for the candy bars he thought he had to have every day.

It has been awhile since I last saw him. I have heard that he is still doing quite well. The final times I saw him, he no longer needed his gloves, and his attitude was much more joyous. He was the kind of person who always tried to give joy to others, so it was a great pleasure to see him enjoy some of it himself. All I did was the very thing I would do for any kind of an imbalance. I did not have to formulate a new product for him. He took the very same things I suggest for any client. Once more, preventative medicine was not too late to produce a great miracle.

Banishing Headache Pain

A pain or ache in the head . . . I have a headache! The headache, it is one of the most common ailments known to humanity. Most of us do not know that it is actually a symptom rather than a disorder in itself. We all are familiar with it. It may come as a stab between the eyes when you eat cold food too quickly. Perhaps it is the ever-tightening clamp of tense muscles around your head. Worst of all, it can be the "lock me in a dark room until it's over" pain of a migraine. If you suffer from a headache, you treat it as most of us do . . . either wait it out or go for the bewildering array of over-the-counter pain medications. Often, this is sufficient treatment. However if you have a headache more than three times a month, you need to take a close look at the possible causes. It is reported that recurrent migraine or tension type headaches are seen in about 50 percent of FM patients.

If you have a headache more than three times a month, you need to take a close look at the possible causes.

There are three types of headaches. The first, a tension headache, is that tightening band of pressure around your head. This is triggered by stress and usually lasts anywhere from thirty minutes to seven days. In the worst cases, it lasts all day every day. Migraines bring on severe, pulsating pain, nausea, and sensitivity to light and noise. Many migraine sufferers retire to a darkened room until the pain subsides. This can take anywhere from two hours to three days. Cluster headaches cause excruciating pain behind or around one eye. Sometimes this pain radiates into the temple, jaw, nose, chin, or teeth. Nasal congestion, sweating, watery eyes, and a flushed face can add to the discomfort. This type of headache may occur three times in a day, lasting from fifteen to ninety minutes for up to sixteen weeks at a time.

Medical doctors urge patients with serious headaches to keep a headache diary. It is very helpful to record the intensity, duration, symptoms, the degree to which the patient is disabled, and how he is affected by any prescribed treatments. The doctors may even recommend fighting headaches with changes in eating, sleep, or work habits. Other treatments used commonly are physical therapy, biofeedback, and drugs. For tension headaches, physical therapy, hot packs, cold packs, or electrical stimulation devices such as ultrasound and massage are the accepted treatments. Biofeedback helps about 80 percent of the time, and it is the only non-pharmaceutical treatment that works for migraines. It is somewhat helpful for tension headaches, but completely ineffective for cluster headaches.

The most common causes of headache are:

- **Allergies**
- **Arthritis**
- **Caffeine withdrawal**
- **Cold or flu**
- **Hangover**
- **Head trauma**
- **High blood pressure**
- **Hunger**

- **Disease**
- **Eye strain**
- **Sinus infection**
- **TMJ**

Foods that cause most of our headaches are aged cheeses, smoked meats, chocolate, alcohol (especially red wines), homemade yeast breads, raisins, avocados, and MSG, an additive in Chinese and processed foods.

Those who suffer from chronic headache are almost always in a state of enervation. Their lifestyles and diets do not work for them, so they experience headache interference. This indicates it is time to make changes. Often it is a change in diet that works the best. The support products used in my practice make healing the headache a simple process. I am thrilled to see the amazing properties of enzymes doing the work for which nature intended them. My experience with headache treatment through the use of enzyme therapy is always the same. The headaches become a thing of the past, even those intense migraines.

When I meet a client for the first time, I cannot always know what causes the headache because many different factors are involved. No matter what the cause, I begin with protease in order to create a healthier immune system. Dosages of 330,000 units of activity several times daily are initially required to make a difference with those who have chronic headaches. This protease action corrects or breaks down the toxins and undigested foods in the bloodstream. One of the products I formulated includes herbs to feed and fortify the nervous system. Other herbs are used as antioxidants that add oxygen to the blood. When these two specific formulas are used together, protease and the herbal compound, the results are stunning. I conclude from this that most headaches are caused by one of two things. First, widespread toxicity in the body, sufficient to prevent the breakdown of food particles in the digestive tract is a primary cause of headaches. Second, a stressful event taking place that involved the nervous or endocrine system. When I give either the protease or the herbs and enzymes combined, the headache frequently disappears within moments. After using this formulation for a short time, those migraine sufferers realize they are not having the headaches as often or as intensely as they once did. If the headaches come

It is important to remember that stress is not an external force. Developing from an internal source, it is the result of cellular contamination and nutritional deficiency.

back, the client must readjust the care he is giving to himself and get right back on track. Keep in mind, a headache is a symptom indicating something is wrong in the body. It is not a disease itself.

It is important to remember that stress is not an external force. Developing from an internal source, it is the result of cellular contamination and nutritional deficiency. Manifestations of stress take the form of pain, fatigue, anxiety, disability, and malfunction. These are the direct results of substandard health conditions. The stress due to poor health can be light, mild, moderate, serious, and severe. If we are experiencing light stress, usually there will not be any kind of physical symptom to inform us of its occurrence. With mild stress we may sometimes feel slightly out of sorts. In moderate stress constant symptoms are present, though they are still sufficiently weak to be merely annoying. Serious or severe stress will undoubtedly manifest in some way, often as strong pain. This is the body reacting to the stress of cellular contamination and nutritional deficiencies. It has finally reached the point where it can no longer endure the stresses imposed upon it without physical manifestations. The common response is, of course, to take some kind of therapy to alleviate the pain and other symptoms. This is certainly appropriate, but it does not address the cause of the problem. It is only a matter of time before the malfunction returns, possibly in the same form. However, it can be in a much more severe form or a totally different one. Looking at it as a healthcare professional, I believe an individual with health stress manifestations should be placed on a health-building program that compliments the immediate therapy prescribed by his or her doctor. This would guarantee that the reasons for the stress are properly addressed.

The new choices offered for a health building program would include foods appropriate for the client's body type, and eliminate the ones that create stress. Plant enzymes will aid the digestion, assimilation, utilization, and elimination of the food ingested. Protease will be taken to destroy cellular contaminants. Other enzyme fortifiers will eradicate nutritional deficiencies. Good health would return to the patient. No matter what the symptom or problem, learn to change your habits for the better and return to a path that gives life to your system.

Stress and Hypertension: Calming Your Nerves . . . Reducing Anxiety

The best example to demonstrate the interdependence of the body's many systems is the nervous system. Earlier in this chapter I characterized the endocrine and nervous systems as the two major control centers. We know the hormones cannot function without a strong and well-balanced endocrine system. Consider that the nerves, like the hormones, are closely related to the endocrine system. One of the physiological responses created by the electrical impulses from the nerves is the secretion of hormones by the endocrine glands. Yet, the nervous system receives its support from the endocrine system. Research shows that people with Fibromyalgia have low growth hormone, which is involved with muscle repair.

The complexity of your nervous system is truly mind-boggling. Electrical impulses in the neuro-chemical process travel throughout your body. The system is a perfectly coordinated network of nerve cells—possibly in the hundreds of billions—that are tied to threadlike formations. These work together to process and respond to information from all your senses. When nerve cells die, they are not replaced.

Hence, an older person has many millions fewer than a youngster. If you suffer nerve damage, only a limited amount of repair is possible. The cells are obviously very sensitive and will not perform well if our brain is perturbed by a concentration of undigested protein fragments or a shortage of oxygen or glucose. All this begins to thicken the plot. Not only is a sufficient supply of enzymes necessary to fortify and revitalize our endocrine system, which in turn, supports our nerves, but those same enzymes are vital for complete digestion. The digestion is what allows our nerves to do their job.

Clients with anxiety problems often complain of other difficulties. In most cases, we can trace these back to the same fundamental cause—an overall bodily imbalance. The lack of internal equilibrium, or challenged homeostasis is the true culprit. After a regimen of enzymes, calming herbs, and the appropriate genetic diet for his or her own specific body type, the client shows a dramatic improvement in both health and mental and emotional outlook.

Hypertension

High blood pressure, or hypertension, can cause or aggravate disorders like heart attack, stroke, and kidney disease. Weight loss will significantly lower blood pressure without the use of drugs. *A Journal of the American Medical Association* study revealed that losing only eight pounds reduces diastolic (the reading taken when the heart muscle relaxes and the chambers fill with blood) blood pressure by an average of 2.3 points. Walking four or five times a week for thirty to forty-five minutes will create an overall reduction of the blood pressure. Eating less salt will lower diastolic blood pressure an average of one point. Diastolic pressure is the second of two numbers in the blood pressure reading such as 137 over 80. Those with a reading of 90 or above have high blood pressure.

The findings of the American Medical Association showed that in spite of the widespread belief that emotional stress can elevate blood pressure, no benefit was gained from participating in a stress management program or taking vitamin supplements. Researchers from eleven medical centers are collaborating on an ongoing project sponsored by the National Heart, Lung and Blood Institute in Bethesda, Maryland. They are seeking ways to prevent high blood pressure from developing in people who face a higher-than-usual risk for it. A diet low in sodium and high in potassium helped, but not nearly so significantly as losing weight.

Body Types Two and Three are the ones most susceptible to hypertension. The reasons are clear; both body types go for strong-flavored, spicy foods with high sodium content. In addition, these types are drawn to fatty proteins. I have discovered that when the proper genetic diet is followed and enzyme deficiencies are replaced, the patient is once again in control of his own blood pressure.

Psychoneuroimmunology

It is difficult to fathom how emotions can trigger illness. George Solomon, M.D., professor of psychiatry at UCLA and adjunct professor of psychiatry at the University of California at San Francisco has demonstrated that emotional stress can affect the immune system. Dr. Solomon is known as the "father of psychoneuroimmunology." This new science is now taught at universities. Psychoneuroimmunology is a field of medicine which joins immunology and neurology. It focuses on the relationship of stress upon the HPA axis (hypothalamus, adrenal, and pituitary). Depletion of hormones and neurotransmitters within the HPA axis, as a result of stress, can lead to a multitude of diseases and disorders. This lengthy word means the study of the effect of emotions on the immune system. More evidence arises every year to support this discipline.

After the decision has been made to create changes in lifestyle, the body responds within a twenty-one day period.

After the decision has been made to create changes in lifestyle, the body responds within a twenty-one day period. If you stop smoking or overeating or taking a health care product it paves the way for the healing process to begin. During the second twenty-one day period, the body begins to balance, adjust to, or accept the changes. In the third twenty-one day period, the body undergoes transformation. This gives the mind an adequate length of time to adjust to the change and continue healing in a positive and beneficial manner. These twenty-one day segments of response to change and therapy are an established regimen among health care practitioners. Although we may feel better immediately when we begin a new program, there is a period of adjustment while our body seeks its comfort zone. Response is heightened in the second period. In the third, the condition or problem that prompted us to take action has definitely improved for the better. This is why my answer to the question about how long we need to be on a program like enzyme therapy is, "Anywhere from six to nine weeks."

If my book has inspired any of my readers to take plant enzyme supplements, they may have to adjust the number of capsules taken with meals during the first twenty-one days. Your goal may be to improve your digestion, or to take a formulation that feeds and fortifies your hormones. For example, a client who is on a set formulation calls to say that the results first experienced have now changed. I will make the appropriate adjustment of the number of capsules taken, to meet this new need. This is a very normal scenario, and one that is frequently experienced. If you have ever tried to stop smoking or eating chocolate, you find the first three to seven days the most difficult. The remainder of the time finds the need or desire much alleviated. This is how and why I advise clients to make periodic adjustments.

Another way to use the science of psychoneuroimmunology is to interview cancer patients or those with other catastrophic diseases. You will be surprised to find that most of them suffered some form of trauma two or more years before the onset of their illnesses. This trauma could be the death of a loved one or surviving a serious automobile accident or any other devastating experience. Some people have actually gotten so involved in their own illness they have precipitated a depressive effect onto the immune system.

Dr. Bernard Siegel has stated on his tapes "Life, Hope and Healing," (Nightingale Conant) that he believes that surviving cancer has as much to do with our mental attitude as it does with the extent of our disease. I definitely agree with him. My own personal experience with cancer proved this to me beyond all doubt. When I finally realized this, I no longer had to blame myself. Rather, I learned my lessons from it and went on to make the positive changes I needed to remain alive, healthy, and happy.

Chapter 5

Revitalize Your Energy Level

Tapping Your Life Force

Are you happy with your energy level? If your answer is "no," you are not alone. Stress caused by career pressures, financial concerns, and relationship tensions can take a major toll on our bodies. This is particularly true for the endocrine and immune systems. These systems sense negative changes in our environment and send a corresponding message to the brain. If these two systems are already weakened by enzyme deficiency, the whole body can feel rundown and lethargic. As we age, the ability to produce sufficient amounts of digestive enzymes decreases. The lack of enzymes results in indigestion and/or malabsorption problems that can interfere with your sympathetic nervous system.

What is the sympathetic nervous system? This part of our nervous system is responsible for crisis intervention. As an immediate response to danger or other traumatic challenges, the sympathetic nervous system puts body processes into high gear. It stimulates secretions from the endocrine glands. It pumps adrenaline into the system, while the adrenal glands themselves are releasing elevated levels of epinephrine into the bloodstream. The blood vessels constrict so sweating will begin in an effort to keep the body cool while under stress. All this working together is what gives the body its ability to flee or fight. The cost for this extra energy is the slowdown of the heart, lung action, and digestion. This is simply another example of the extraordinary functions our bodies are capable of performing.

What is the endocrine system? The endocrine system is a group of glands located in various parts of the body. Each gland has one or more specific functions. They are all dependent on one another for maintenance of normal hormone balance in the body. As hormones are secreted from these glands they are delivered directly into the blood. Many people refer to them as the system that controls the entire body.

Many clients come to my office with conditions of fatigue and an overall feeling of un-wellness. After the usual interviews and tests and a close look at their medical histories, I recommend a regimen of enzymes for them. The enzymes will rebuild the organs and tissues and strengthen all the processes and systems that keep their bodies strong, refreshed and balanced. After a few weeks have passed, I ask the key question: "Do you feel good?" If the answer is "yes," I know the enzymes are doing their job. At this point, I suggest to the client a reduction in the number of daily supplements he takes, as long as his good health and equilibrium remain stable.

Insomnia

Scientific research shows that those suffering sleep deprivation, or insomnia, have some specific symptoms that are not experienced by normal sleepers. These are higher rectal temperatures, higher skin resistance, more constricting of the blood vessels per minute, and more body movements per hour. These characteristics are fine and are advantages in many different arenas in the workplace, but they are not conducive to good sleep.

The causes of insomnia may be physical or psychological or both. During sleep the metabolic rate should decrease approximately ten percent below daytime levels. This drop is due to muscle relaxation and reduced activity of the sympathetic nervous system. The less relaxed your muscles are, the greater your metabolic rate. Any emotional strain can cause increased muscle tension with a corresponding upswing in metabolic rate. Personal problems and their attendant emotional stress, or other anxieties can create a domino effect. The less we sleep, the more our bodies are susceptible to illness. Added to this we have a new anxiety to worry about—our inability to sleep. Insomnia can adversely affect not only our personal lives and our ability to do our work, but also our bodies' healing capabilities and energy levels.

Some of the physical causes of insomnia are easily solved by changing eating and sleeping habits. Caffeine drinks like coffee, tea, and cola speed up the metabolism, making it more difficult to sleep. Sugars have the same effect. Sometimes a softer or firmer mattress and pillows can allow for a more restful night's sleep. Avoiding heavy meals just before bed may also reduce the possibility of insomnia. If you have fibromyalgia, one type of dysfunctional sleep is called the alpha-delta sleep anomaly. As soon as people with FM reach the deep level sleep, alpha brain waves invade and jolt them back to shallow sleep. Not only are they denied refreshing sleep, but the delta level is when the body does its repair work, chemical replenishment, and the making and use of the growth hormone.

If you have made these adjustments to no avail, I suggest plant enzyme supplementation. This supports your nervous system in general, but now we want to target the sympathetic nervous system. Remember, this is the part of the nervous system that creates increases in energy. The cost for this is the inability to relax into a restful rather than a restless state. Insomnia is another imbalance that has far-reaching effects. When we do not get a proper amount of sleep for our body, it cannot heal itself. One of the reasons why the elderly often cannot sleep well is their bodies are becoming more and more enzyme deficient.

Chronic Fatigue Syndrome

Chronic Fatigue Syndrome, or CFS, has been referred to as chronic Eppstein-Barr virus, chronic mononucleosis, and Icelandic Disease, to name a few. Symptoms include fatigue, nausea, vomiting, abdominal cramping, muscle pain, joint aches, and lack of concentration. Millions of people suffer from one or more of these symptoms. Reports have indicated that more than five million persons in the United States have CFS. CFS cannot be diagnosed by standard tests, like urinalysis or blood work. Some medical personnel still will not acknowledge that it even exists. Yet they are unable to explain the fatigue or illness that many people describe. Countless books have been written about Chronic Fatigue Syndrome. In every one of them, there is no shortage of radical diets to follow, herb mixtures, nutrients, and support groups.

Case History

I have before me many charts with many similarities. However, Chronic Fatigue Syndrome manifests itself quite uniquely with each case that is documented. Four years ago, one of my clients came into my office with CFS symptoms. Here is her description of how she felt in her own words. "I haven't felt well for the last eight months. I was okay for about a year before this, but now I am not able to hold down a regular job. My eyes feel tight and burn and when they water they sting. I have pressure in the head area and I can't seem to think or function. My throat swells and feels like it is going to close sometimes. I have constant sore throats and my lymph glands swell in my throat and underarms. My underarms sting when I perspire. Hives are brought on by heat or when I scratch myself. I tend to run fevers. My joints and muscles ache, sometimes very badly. I have trouble sleeping and wake up tired. I have pains in my chest, in fact when I get up my right side feels like it is in a knot or swollen. I have swelling in the liver and gallbladder area and sometimes you can even see it down at the bottom of my ribs. My feet ache most of the time and my hands go numb. I get chilled easily. I go from constipation to diarrhea. It is very easy for me to get short of breath. I have had kidney and bladder infections off and on. I experience PMS and cramps whether I am in my period or not. I have nausea, depression, and ringing in my ears." She was frightened and discouraged. Various tests were performed that clearly showed her to be a textbook case. Realizing she had CFS, however, did nothing to help her recover. She tried every treatment known but they made no difference.

Did she have CFS? It had not been diagnosed by a medical doctor, but she certainly had the symptoms. I promised her I would not lie to her and tell her if she took enzymes she would recover immediately. However, if she would follow my suggestions and begin taking plant enzymes, I felt certain she would be back on the long road to better health. How would she know this? She would feel the difference in a few days, especially with the nausea. I gave her high potency digestive enzymes with her meals. She was on 330,000 units of protease five times a day, including between meals. Within the first twenty-one days she felt a marked improvement in controlling her headaches, nausea, vomiting, sore throat, muscle pains, and joint aches. The second twenty-one day period she was working more hours per week at her job, and her diarrhea and constipation problems became balanced. Her chest pains disappeared and the swelling in her lymph nodes had noticeably reduced. Within nine weeks she was working full-time again. To this day she is doing very well and living a full life. She cannot eat any junk food or drink alcohol without suffering ill effects, but she has gained control of her life. I treated her case over four years ago, and this young woman still uses enzymes and stays well.

On Thursday, July 17, 1997, Dr. Robert Suhadolnik and his research team at Temple University School of Medicine, Philadelphia, PA, reported that studies of patients with Chronic Fatigue Syndrome have led to the identification of a new human enzyme.

This new enzyme has lower molecular weight than the normal enzyme found in the viral pathway in which this protein is active and may explain common observations in patients with CFS. CFS patients have an inability to control common viruses and an inability to maintain cellular energy. According to Dr. Suhadolnik, the viral pathway known as the 2–5A Synthetase/RNase L antiviral pathway, may control both processes. He further stated that "This new enzyme in CFS may not function as well as the normal RNase L found in healthy people. It may explain why CFS patients' bodies have a hard time maintaining the energy necessary for cellular growth."

As with any autoimmune disorder, dysfunction, or imbalance, I give the same regimen of enzymes, supported by a high dosage of protease. I believe in feeding and fortifying the endocrine and nervous systems and detoxifying the body. With enzymes all these things are possible. The endocrine system (glands that control all hormonal secretions), the nervous system, and the immune system must be addressed. Because they all function on enzymatic action, I know of no better way to support them than through the use of enzyme supplements. Plant protease has a wide range of proteolytic activity. This means the body uses it as needed.

How to Boost Your Brain Activity

Fatigue can be mild in some clients and yet incapacitating in others. The fatigue has often been described as "brain fatigue" in which clients are totally drained of energy. Sometimes they are too tired to remember, let alone function, in the moment. They tell me their brains just are not functioning. Healthy brain activity requires healthy nerves, a powerful endocrine system, and a steady supply of oxygen and glucose. These requirements depend on a properly functioning digestive system. As we age, our body produces fewer hormones and chemicals critical for memory function. While memory loss in the elderly can be attributed to other sources, proper enzyme supplementation can help to fortify and restore the efficient operation of major organs. Insufficient exercise and inappropriate combinations of medication do cause memory loss in some of our senior citizens. However, maintaining a proper balance of enzymes in our bodies can renew production of memory-enhancing substances. The results can be very impressive. Memory loss can be significantly reduced, and often halted altogether. The big lesson here is that it is much easier to prevent age-related memory loss than it is to restore it.

Enzymes Strengthen and Repair Your Muscles

Enzymes may be the most important, but also the most neglected, essential element of vibrant health. In fact, enzymes may be what makes the difference between our fit, young, Olympic athletes and those youth who choose a path of sluggishness. Body builders have a special appreciation of enzymes. If muscle tissue enzymes were not at work there would be no growth of new muscles. No neurological responses would occur to make the muscle work at all. Enzymes are the catalyst that transforms food into the necessary energy to make muscles move and grow. Because oxygen is one of the sources of enzyme production, proper breathing and exercise are very important for enzyme metabolism.

Natural exercise, like walking, helps generate metabolic enzymes. If you are not doing regular, healthy exercise, you probably have an enzyme deficiency. Creation of enzymes is diminished with disuse of the muscular system. I am not referring to strenuous exercise or weight lifting. I favor exercise that is natural to the human body, like walking, swimming, or bicycling. Scientists are now showing a greater interest in nutrition than ever before. Still, very few have discovered the most important and invigorating factor in a truly potent diet—enzymes. We are all our own living, breathing pharmacies. Each of us carries a built-in fountain of youth. To turn it on and keep it perpetually running, do what it takes to step up production of metabolic enzymes.

Each of us carries a built-in fountain of youth. To turn it on and keep it perpetually running, do what it takes to step up production of metabolic enzymes.

The feeding of muscle cells requires forty-five nutrients. These must be delivered to those cells by enzymes. If you desire to build muscles or take part in an active fitness program beyond natural exercise, these forty-five essential nutrients must be provided in an even greater quantity. If these nutritional components that increase stamina are not created, disastrous results may ensue. Many people have attempted to build their bodies and have experienced nothing but greater fatigue. This is an indication that you are lacking enzymes in your system. Enzymes occur naturally in every living cell. Without them, there would be no life. We must find a way to continually refurbish our bodies.

We have ample proof that we loose enzymes in our perspiration. I have analyzed the blood of many athletes. Those who run or constantly "over do" tend to have some of the worst blood I have ever seen. Exercise helps with the removal of waste from our system. That is, a moderate, healthy amount of exercise works in our favor. There is such a thing as over-exercising in which the body begins to break itself down. The forms of exercise that help to remove toxins are brisk walking several times per week or perhaps a cardiovascular workout at a local club three times per week. These generate metabolic enzymes. Be cautious and always strive for balance in all facets of your life.

Enzymes transfer the energy resulting from good, complete digestion to the muscles, nerves, bones, and glands. This helps with the reduction of swelling and inflammation. These "wonder drugs of nature" build muscle and enhance muscle movement and coordination. They maintain sharp memory, mental well-being, and physiological stability. They afford us protection from the hazards of environmental pollution and other toxins to which we are constantly exposed. Without enzymes, we could not exhale carbon dioxide when we breathe. Any interruption or corruption of these vital functions would be disastrous. Imagine how our health would suffer if we could not produce energy or build and repair tissue and muscle. We would be rendered helpless and our bodies would decay. Many diseases attack us when we lose these enzymes. It is a known fact that if we exercise or exert our bodies beyond the fuel we have available to us, catabolism sets in. This breaks down muscle and tissue, leaving us in a far worse position than we were before we began our exercise program.

If your goal is to increase muscle strength, beware of too many high protein drinks and high carbohydrate energy bars. Because these products contain large quantities of sugars along with quickly absorbed proteins, the circulatory system becomes sticky due to the excess protein in the bloodstream. This stickiness keeps the blood cells from delivering oxygen to the tissues. To remedy this, supplemental enzymes will help with the digestion of these drinks and bars. High dosages of protease between meals will break down the excessive protein content in the blood.

Only 20 percent of FMS cases have a known triggering event that initiates the first obvious "flare." During a flare, current symptoms become more intense, and new symptoms frequently develop.

Myofascia

Myofascia is a thin almost translucent film that wraps around muscle tissue. It is the tissue that holds all the other parts of the body together. It gives you shape and supports all of the body's musculature. You can see myofascia if you cut up a fresh chicken. It is the thin, sticky, somewhat filmy material that wraps around the muscle tissue. It wraps around muscle fibers, bundles of fibers, and the muscles themselves, and then goes on to form tendons and ligaments. For people with fibromyalgia syndrome (FMS) and/or myofascial pain syndrome (MPS), the myofascia takes on a new importance. Tightening and thickening of the myofascia occurs in many cases of FMS and/or MPS. If both of these conditions are present, this tightening causes more than double the trouble. When the myofascial tissues become thickened and lose their elasticity, the neurotransmitter's ability to send and receive messages between the mind and body is damaged, and the communication between the mind and body is disrupted. Myofascia, then, may well be the key to what is wrong with people

FMS/MPS. In the myofascia there is a material called ground substance. This material can exist in a solid, semisolid, or fluid state. When ground substance changes from a liquid to a gel, the myofascia tightens, and it is difficult to get it to reverse to a liquid state again without intervention.

Myofascial Trigger Points

Trigger Points (TrPs) are found as extremely sore points occurring in ropy bands throughout the body. They can also be felt as painful lumps of hardened fascia. The bands are often easier to feel along the arms and legs. If you stretch your muscle about 2/3 of the way out you might be able to feel them. Sometimes the muscles get so tight that you can't feel the lumps, or even the tight bands. Your muscle feels like "hardened concrete." TrPs can occur in the myofascia, skin, ligaments, bone lining, and other tissues. They can be caused by a surgical incision, as is often the case with abdominal surgery. You have probably never heard of TrPs, yet they are quite common. Each specific TrP on the body has a referred pain or other symptom pattern that is carefully documented in the Trigger Point Manuals.

The first time I opened the Trigger Point Manuals ("Myofascial Pain and Dysfunction: The Trigger Point Manual Vol. I & II" by Travell M.D. and David Simmons M.D.), I was dumbfounded. After being told for so many years by medical experts that the pain patterns I described did not and could not exist, seeing them illustrated in a medical text brought a flood of emotions. I felt so relieved I cried. I felt validated. Then, as the truth started to hit home, I started to get angry. Why didn't these "experts" have knowledge of Travell and Simmons' work? Why hadn't I learned about these texts in medical school! Most specific pains commonly attributed to FMS are actually from trigger points. TrPs seem to form throughout life as a response to many things that happen to our bodies. Overuse, repetitive motion trauma, bruises, strains, joint problems, etc. Pain creates a neuromuscular response, and the muscle around the pain site tightens, "guarding" the hurt area.

When muscles are in a state of sustained tension, they are working, even if you are not. A working muscle needs more nutrition and oxygen, and produces more waste, than a muscle at rest. This creates an area in the myofascia starved for food and oxygen, and loaded with toxic waste—a trigger point.

Dr. Janet Travell, in her autobiography *Office Hours Day and Night,* explains how dizziness, ringing of the ears, loss of balance, and other symptoms can all be caused by TrPs in the side of the neck, in the muscle group called the sternocleidomastoid (SCM) complex. This muscle has many functions, one of which is to hold your head up. Receptors in the SCM complex transmit nerve impulses, inform the brain of the position of the head and body in the surrounding space. With TrPs, the receptors lies. What they tell the brain is not what the eyes tell the brain.

Developing secondary and satellite TrPs can give the false impression that the MPS is a condition that will steadily worsen with time—that it is progressive. MPS is not progressive. With the proper intervention, these trigger points can be broken up and eliminated.

FMS is, among other things, a systematic neurotransmitter dysregulation, with many biochemical causes. There are other problems as well, but they are all systematic in nature, such as the alpha-delta sleep anomaly. Myofascial Pain Syndrome, however, is a neuromuscular condition. MPS happens because of mechanical failures—the mechanics of physics, not biochemistry. Due to the nature of trigger points, some of the symptoms may seem to be systemic, but they are not. Initiating events, such as repetitive motion injury, trauma, and illness, can start a cascade of TrPs.

FMS/MPS Complex

People with the FMS/MPS Complex face more than just two sets of symptoms. Today, researchers are realizing that FMS and MPS not only occur together, they reinforce each other. Therefore, physical therapy and all other forms of treatment must proceed carefully. Any treatment regimen will be more complicated and less successful than if the patient had only one of the two conditions.

In FMS/MPS, a chronic pain condition exists, with many different symptoms. The trigger points of MPS are all magnified. The body's requirement for protease and lipase enzymes is greatly intensified. The necessity for enzyme therapy is essential for restoration.

Chapter 6

Pamper Your Sexuality

Revitalizing Your Endocrine System

What in the world is the endocrine system? I will give you a very simplified version of the answer to this question. Anytime a full medical description is given of something as complex as a part of the human body, it loses something in translation. The last thing I want to lose is the interest and attention of my readers! In short, your endocrine system is a family of glands that provide hormones for balance in the body. The major members of this gland family are the thyroid, parathyroid, pituitary, adrenals, the alpha and beta cells of the pancreas, and the gonads. The gonads are the sexual glands; testes in the male and ovaries in the female. Your endocrine system is one of only two major control systems in the body. The other one is the nervous system. They are very closely interrelated.

So what do glands do? Glands secrete hormones into your circulatory system, each of which is intended for a specific site within the body. These hormones control the tissues or other glands they reach by regulating the rates of chemical reactions inside the cells. Hormones have a profound effect on important functions such as growth, reproduction, and the elimination of toxins. These chemical reactions must be precisely orchestrated so that the targeted glands or tissues perform their tasks at the exact level of activity. If a healthy medium is not preserved, the originating gland or the target gland can be damaged. Some of the major illnesses that plague us arise when too few or too many hormones are released. An example of this hormonal imbalance is menopause. When this stage of a woman's life arrives, the ovaries cease to produce estrogen. From that point on, she must depend on the adrenals to take up the production for her. Another situation could be when a client is unable to maintain a balance of calcium in his system. Calcium levels are regulated by the thyroid, along with the rate of metabolism. If the thyroid has been undermined, our ability to control our weight and our sleep patterns is adversely affected. When any of these situations take place, we are said to be suffering from a hormonal imbalance.

Hormone Imbalance

A hormone is a substance formed in one gland and carried in the bloodstream to another gland or tissue where it serves a specific purpose. Although the total number of hormones the body produces is unknown, we do know that each hormone has its own unique chemical composition. Each gland of the endocrine system manufactures several different hormones. The adrenal gland alone produces more than twenty-five varieties of hormones. Growth, reproduction, sexual attributes, and personality traits are all dependent on hormones. Insulin is a well-known protein hormone that regulates carbohydrate metabolism. It is produced by the beta glands in the pancreas. Dr. William Regelson and Carol Colman do such a wonderful job describing the needs of hormone balance that I won't go into great detail in this book. I suggest reading *The Super Hormone Promise* (Simon & Schuster). In it, Dr. Regelson provides complete and comprehensive information about each of the super hormones, including up-to-the-last minute research on melatonin, DHEA, pregnenolone.

If we want our bodies to function efficiently, our hormones must be in correct balance. Hormonal imbalances can cause problems ranging from lethargy to excitability. We now know that any significant imbalance in the kind and number of hormones generated must be corrected if our mind and bodies are to be in a good state of health.

Premenstrual Syndrome (PMS)

In some cultures, the female cycle is considered an important time for women. It symbolizes a woman's power to create life. However, if you are a woman who suffers from Premenstrual Syndrome (PMS), you may not feel quite so powerful. While PMS manifests itself differently in different individuals, the most common physical symptoms are headaches, breast tenderness, weight gain, and sugar cravings. Some of the psychological effects include mood swings, depression, irritability, fatigue, and tension. There have been one hundred fifty symptoms of PMS reported. Many causes have been proposed for it: Estrogen and progesterone imbalance, how the body metabolizes glucose, a deficiency of the brain chemical serotonin, hypoglycemia, and candida. Those women who are most likely to develop PMS are those who suffer from sugar imbalance and intolerance.

I suggest many different preventive methods for dealing with PMS. Women suffering from PMS should avoid caffeine, excessive salt, alcohol, and smoking.

Abstention from these toxins should always be observed, not just before or during the menstrual period. Increasing your intake of complex carbohydrates, protein, and exercise help considerably in the management of stress due to PMS. A good thirty minute walk every day does wonders for reducing susceptibility. Enzymes can be used as a preventative for PMS. I advise women to increase their protease dosage ten days before their period and make some diet changes as well. Protease should be taken in the mornings, evenings, and midafternoons. A digestive formulation high in amylase (25,000 DU) will help sugar cravings to subside.

One of the questions appearing on my health questionnaire asks if you have suffered, or do you have, PMS. Almost all women answer "yes" to this question. PMS is regarded as a disorder or symptom that makes you feel unwell during the time of menses. Is there a way we can prevent this, before we ever experience the first symptom? The answer is "yes" and the preventative is enzymes. I think it is wonderful that education and information about enzymes is becoming much more widespread than it was several years ago. Since 1986 I have seen young parents starting their children on an enzyme program at early ages. Young girls who have been on enzymes have a significantly lesser occurrence of PMS than others. Even after symptoms are observed, something can be done to alleviate the chaos caused in the body from PMS.

Case History

A sixteen-year-old girl came to see me. About two weeks of her month are fairly normal. But the other two find her with mood swings, cramps, and a good case of the "uglies," as she puts it. Her mother had taken her to several OB/GYN doctors and nothing worked except heavy doses of prescription drugs. The mother did not want her daughter to have to endure her teenage years in a drugged stupor, although her father thought that might be a very good idea! I suggested the basic regimen of enzymes, but reminded the girl it was her responsibility to follow the directions. I asked her to make another appointment with me after she had two more periods, during which time she would be using the enzymes.

She reported her first period under the enzyme program seemed easier. Instead of her usual bloating and depression, her period started before she even realized it. She was sure that "It was all in my head." With the second cycle there were no cramps, bloating, or depression. I modified her enzyme dosage and explained to her that she would no longer have mood swings if she did away with soft drinks and candy bars. I told her to eat more protein foods. She called to cancel her appointment that had been scheduled right after her third period. She said she did not need to come in, that all her symptoms were gone. By learning which foods triggered her anger and other symptoms, she has eliminated them from her diet. Coupled with an increase in her protease, she feels great.

Premenstrual symptoms are a reality. Many women suffer from them. I encourage you to do what you can to relieve the symptoms. Often, the solution is to educate yourself and find out what works best for you. When you have a strong sugar craving, this is your body's way of telling you it needs nutrients or it wants support. There is a big difference between this and how we mistakenly think our body needs sugar. The more sugar you eat, the more the brain cries out for nutrients, and you think that means more sugar. Hence, the roller coaster ride never comes to an end. When you fortify your body and give it what it needs, the cravings disappear. Taking amylase, the enzyme that breaks down all the sugars you crave and eat, will assist you in overcoming irresistible temptations. If you continue to support your system with enough amylase, you will discover that even you, yes, you can give up sugar. I recommend the dosage for protease to be 375,000 units of activity taken at one time. For PMS, you might want to try 31,500 units of amylase activity to dispel the yearning for sugars.

Menopause

Menopause, or the "Change of Life" occurs when a woman's menstrual cycle wanes and gradually stops. Along with this, the ovaries stop functioning. This natural process results from the normal aging of the ovaries and occurs when they can no longer generate ovulation and estrogen production. While the first and most obvious sign of menopause is the change in menstrual flow, the gradual cessation of estrogen production causes many physiological changes. The fallopian tubes, uterus, cervix, and vagina all become smaller. The average American woman will experience menopause between the ages of forty-seven and fifty-one. Almost 75 percent of women will begin it in their forties. In some instances, it can begin as early as thirty-five or start as late as fifty-five.

The length of time it takes to complete menopause varies, but it usually lasts anywhere from six months to three years. Most women pass right through menopause with little discomfort. Only about fifteen percent experience discomfort and distress. Occasionally during menopause, some existing physical ailments may become exaggerated. Hot flashes, one of the most common symptoms of menopause, are sensations of heat in the face and upper body.

There is a lot of information available on DHEA, melatonin, and the youth hormones, but virtually nothing on the enzymes required for these hormones to work.

Often they are followed by sweating or chills. Some women feel just a few of these hot flashes over a period of time, while others can experience ten to twenty per day. Those women who do suffer psychological reactions may experience fatigue, crying spells, insomnia, an inability to concentrate, or lapses in memory. More severe reactions may result in depression. At one time it was thought that the changes in the body's chemistry caused these reactions. Today, many physicians believe that the chemical changes which bring on menopause simply triggered reactions to other events that occur at this time of life. We need to be aware that as we age we loose digestive enzymes along with the enzymes that produce DHEA. There is a lot of information available on DHEA, melatonin and the youth hormones but virtually nothing on the enzymes required for these hormones to work.

Case History

Menopausal women who have used enzyme therapy report that their hot flashes and night sweats are things of the past. One forty-nine-year-old woman was experiencing fatigue, hair loss, hot flashes, depression, and insomnia. Although she tried over-the-counter menopausal remedies, she found no relief from these serious symptoms. As soon as she began treating her entire body, especially her endocrine and nervous systems, with enzyme therapy, her condition began to improve. Her hair was healthy, her depression and fatigue lifted, and she began to sleep at night, free of hot flashes.

In other cultures and in times past, a woman who gave up her blood, (menopause) could at last hold in her own energy and power. Known as crones, they were regarded as the most powerful and wisest of women. Within a tribe, clan or village, they were highly valued and revered. Nowadays, many women in our culture live to be over eighty years old. A full one-third of these women's lives are lived after menopause has come and gone. It should not be looked upon as a curse or an illness, anymore than menstrual periods. It is not an enemy.

Surgical Menopause

If a woman has her ovaries surgically removed, she experiences menopause immediately. This is because her estrogen production has been terminated. This menopause can be more traumatic because the body was not given time to gradually accept the loss of estrogen production. Enzyme therapy can make this process less emotionally disturbing by supporting the endocrine system. The enzymes aid in helping the body to adjust to the estrogen loss. Eventually it will resume production of its own metabolic enzymes.

Hormonal Replacement Therapy (HRT)

Should you or shouldn't you? Hormonal Replacement Therapy, or HRT, involves taking synthetic estrogen when the ovaries no longer produce it. HRT is certainly not for all women. The decision to take it must be very much of an individual one. There have been many findings and recommendations published from all the research done on this controversial topic. Estrogen replacement was introduced in Germany in 1896 where ovarian extracts were used. Today, in our own age of chemicals, estrogen is commercially prepared. Synthetic estrogen is the most commonly used medication on today's market. In the fifties and sixties, estrogen therapy was touted as "feminine forever." During the seventies, some studies indicated that the use of synthetic estrogen had cancer-causing effects on the lining of the uterus. Nineteen eighty and the ensuing decade told us if we added progesterone, the harmful side-effects of HRT would be easier to control. In the nineties, we are questioning the use of chemicals, pills, patches, creams, injections, implants, and sublingual (under-the-tongue) therapies. Countless books on the subject denote a bewildering array of information. At first glance, it would seem that estrogen replacement is the answer. However, each book I have read ended in doctors' cautioning it was not for everyone. No wonder there are so many confused women struggling with this decision that is critical to their health.

I personally chose to use natural ingredients delivered by enzymes, even though I had a hysterectomy at the age of twenty-eight. These super-hormones involve pregnenolone and DHEA delivered by lipases. I would never tell anyone what to do. This happens to be my personal choice and it works for me. I am now in my fifties and feel very healthy and vibrant. Again, it must be a personal choice. I believe the important word here is choice. What are your choices?

Current reports indicate that less than 20 percent of menopausal women choose hormonal replacement therapy. The decision about this treatment is highly personal, and depends on a woman's family history, personal habits, and the risk factor she finds the most tolerable. This is not a one-size-fits-all situation. In my work, the question arises, "Can the woman who is a longtime follower of natural health approaches skip HRT without damaging her health? Dr. Regelson describes the new "designer" estrogens that are being offered. The estrogen needs are different from woman to woman.

The human body is a great marvel. The endocrine glands secrete vital hormones which control many bodily functions. Everything is designed to be perfectly synchronized when optimum health prevails. If hormone production is less than adequate, it is because the once reliable glandular systems are unbalanced and weak. One of the miracles of our natural design is the manner in which the body compensates for lost or damaged components. For example, when the ovaries have been surgically removed or cease to function because of menopause, other glands fulfill the need for estrogen. These glands release hormones that perform all the functions of estrogen except sustaining the menstrual cycle. These glands are also a part of the endocrine system.

This raises another question: Why do we need estrogen replacement or why do we have PMS if other glands produce hormones that seem to replace estrogen? When the organs and systems of the body are in a depressed state, we cannot expect it to do everything it is programmed to do without some external support. This is why women often have to opt for HRT. In fact it is the enzyme that assists in the making of the hormone that stops aging. If you are missing this enzyme, the whole hormonal cascade unwinds.

Our endocrine system is often referred to as the "body mind." This will give you a good feel for how vitally important it is. It virtually runs the systems of the body. The hypothalamus in the brain sends signals to the pituitary—the master gland. The pituitary is responsible for distributing this information to all other glands by way of the hormones moving in the bloodstream. This can be likened to a fine quality, state-of-the-art, high tech piece of electronic equipment. Because it is so fine-tuned, no electrical shorts can occur. However, if the endocrine system is not running at maximum efficiency, it malfunctions. It can no longer meet the demands of the body. This system runs on enzymatic action, as does the nervous system. When the glands are undernourished, it takes enzymes to deliver the nutrients from the food or vitamin and mineral supplements. Without enzymes, these nutrients cannot be assimilated.

The menopausal women I have counseled are in the thousands. They have many things in common. The one ailment we can eradicate completely is hot flashes and the attendant night sweats. These symptoms do not stop immediately, however with enzyme therapy, combined with correct foods for an individuals body type, it only takes a matter of weeks before they dissipate. This usually takes twenty-one days. I have known some women who have experienced remarkable changes in only three days.

Case Histories

A fifty-one-year-old woman came to me for nutritional advice. She did not mention that she was having severe hot flashes, because she believed everyone had them. They came with the territory. I helped her determine her body type, and she proceeded with her enzyme therapy. I did not give her the extra enzyme mixture I normally give to menopausal women. I did not ask, and she did not volunteer any information. After three days, she called to ask if she was hallucinating. She was amazed that her hot flashes had drastically decreased. It is very rewarding for me to see these changes take place in my clients, especially when there are no expectations involved. Many women survive PMS and its rigors, and for the same reason. When the body begins to balance, it heals itself. When the glands are well nourished, the body will not suffer from malfunction or malnutrition.

Another lady came to see me in a state of great anxiety. She was forty-nine. Her face was breaking out, she was loosing her hair, she would wake at two o'clock in the morning, unable to go back to sleep; she was exhausted. None of the menopausal aid products she obtained from the health food store had worked for her. She could not continue, her life was not working. Whatever she tried, relief would not come to her. I told her that she had to feed and fortify her body, and treat it with great love. Although she had taken the creams mixed with different herbs to make progesterone, there had to be a more complete balance. I knew her endocrine system had been badly undermined. When a woman enters menopause, she can no longer count on her ovaries for hormonal support. She must rely on her adrenal, her thyroid and the other glands of the endocrine system. This woman, my client, finally understood that she could not get by on over-the-counter quick fixes. She had to treat her body as an interdependent organism. When she began using the enzymes, she made tremendous progress in only twenty-one days.

I insist that the people who come to me for counseling commit to a program for at least six weeks, but preferably nine weeks. The ideal length of time is nine weeks. The lady whose case I just reviewed for you had become so deficient, with a very weak immune system that I kept her on quite an intense dosage for four or five months. But what works for one does not work for all. The enzymes always work, but the time frame can vary greatly, depending on an individual's needs. Her need was greater than in the first woman's case. The first step is always to determine where you are in your own healing or enervation process. That way, you can decide how long you need to treat and feed yourself. To expect to turn your state of health around in only three days is unrealistic. I would encourage all women who are entering menopause to give themselves time to heal. Remember, educating yourself is the key; then make choices that fit your lifestyle.

The enzymes always work, but the time frame can vary greatly, depending on an individual's needs.

Different body types respond differently to estrogen therapy. The Neuro Body types, those who are underdeveloped or boyish, respond well to it. This is because their estrogen levels are low in the first place. The Supra Body types, or overly developed women, however, do not respond as well. These women tend to have larger breasts and experience blood sugar problems. The Supras already have high estrogen, as do the Estro Body type. Both Supras and Estros seem to be irritated by estrogen replacement therapy. For these latter two types, I make sure they feed and fortify their entire endocrine system. I would never ask a client to stop taking estrogen if her personal physician had prescribed it for her.

Those women who do not do well with HRT have a connection with the way they are built (genetics) and with some of their food cravings. Some of the symptoms that may occur while on estrogen are fatigue, depression, moodiness, bloating, headaches, or irritable bowel movements. I suggest ways to support these problem areas so the client can make her own choice about whether or not she wants to continue with estrogen or try another therapy. One of these choices is wild yam cream with progesterone, which can be purchased from health food stores, some clinical nutritionists, or other health-care professionals. For a woman who wants to take the natural path, some suggestions are: Enzyme therapy, superhormones, pregnenolone, DHEA and lipase, vitamins and minerals from natural herbs, or herbs glandular specific to the endocrine glands, a body type food plan that specifically suits you, extra protein for insulin control, stress management, powerful thinking, attitude, relaxation, expressing your feelings and needs in an appropriate manner, living in the moment, laughter, visualizing yourself as you want to be, and exercise.

It needs to be noted that there is a sequential order in the hormonal balancing system. This order is as follows: Good insulin balance which in turn affects healthy DHEA balance, which in turn affects every hormone in the body. This entire sequential order is directly affected by enzymes. The precursor to DHEA is the hormone pregnenolone, which ceases to be produced when enzyme production stops. This situation raises the age-old "chicken and egg" question: Do we age because our enzyme production stops or does enzyme production stop because we age?

Prostate Problems

The American Cancer Society estimates that over 344,000 men will be diagnosed with prostate cancer in 1999. Forty-four thousand of them will ultimately die because of it. Researchers do not know what causes prostate cancer, although it occurs more frequently in older men, men with a family history of cancer, and Hispanic and African-American men. If detected early, prostate cancer can be treated with radiation, drugs or surgery. From my own experience, I have observed that most men who have prostate problems enjoy fatty or salty foods, and may also have a sweet tooth.

Swollen prostates do not necessarily mean cancer. However, if swelling is involved, a medical doctor will closely monitor this patient to watch for signs of the onset of cancer. Men who are vegetarians tend to have less susceptibility to prostate cancer. If the prostate is swollen, painful and frequent urination is a symptom. After beginning enzyme therapy, this discomfort will dissipate. Doctors have found that men using this therapy experience reduced swelling in their prostates. I have them take concentrated dosages of protease and amylase. Protease in massive doses does cause frequent urination, but it relieves the pain caused by the swelling. As in the case of the woman who did not bother to mention her hot flashes, many of my male clients will mention, after the fact, how much better their prostates feel after regular enzyme therapy.

Case History

A fifty-five-year-old man with an enlarged prostate began enzyme therapy under my guidance. He added some necessary diet changes to his program as well. His doctor decided not to proceed with the scheduled surgery that had been planned. Instead, the man had begun eating raw foods like salads, fruits, and vegetables, and had cut back on his intake of spicy foods. He eliminated red meat from his diet but continued eating fowl. He took high dosages of protease in the morning, midmorning, midafternoon, and evening. The swelling of his prostate went down along with the urinary discomfort. He is doing beautifully and has recommended that several of his friends with similar problems come to see me for nutritional counseling.

Balancing and Stabilizing Your Hormones

Hormones are very powerful substances, and extremely small amounts are sufficient to accomplish the work for which they are intended. These chemical messengers are released into the bloodstream by endocrine glands. Their mission is to travel through the bloodstream and lymphatic systems as regulators of the physiological activities of cells or organs. When the hormones arrive at their destination, they act as stimulants for certain receptor cells. The ultimate goal is to create and maintain homeostasis . . . internal equilibrium . . . a balanced body.

There are specific hormones such as estrogen, cortisone, and insulin that show us beyond doubt how interrelated the endocrine system is with the rest of the body. These miracle substances have amazing powers, effecting dramatic changes in everything from pain control to the sex drive to mood swings. Imbalance can occur when certain tissues and glands become too active or too inactive because the hormone support released from a regulatory gland was either insufficient or excessive. To maintain a healthy body, you must be sure your glands are getting enough protein. But that is not enough. It must be usable protein. This protein must be of a kind that will be metabolized into useful amino acids. The best way to obtain these proteins is through a proper diet and appropriate enzyme supplements so that the body can correctly digest and deliver these proteins. This is certainly where body types become important. When you know your body type, you can determine the best sources of protein for you to eat. The wrong proteins can interfere with the acid/alkaline balance necessary for good health.

If every action, including hormonal action, is run by enzymatic activity, does it not make a great deal of sense to keep the body fortified with a continuous supply of enzymes?

It is not unusual for women to come in for counseling and tell me they have already experienced early menopause. I am very honest with them when I warn them, "Please understand that the body may be in a state of imbalance and when we begin to give it what it needs, like life-giving enzymes, it is normal for those in 'early menopause' to begin having periods again." This is met with a variety of responses.

Case History

One young married woman who had not experienced regular periods for years came to my office. She had given up on becoming a mother; at the very least she wanted a healthier life. Several doctors she had seen throughout her married life told her she was unable to conceive. She had not felt well for a long time, and wanted something to boost her energy level. Enzyme therapy proved to be exactly what she needed. Her periods started again and were very regular. Within a few months she was pregnant. I did not promise her or give her hope that pregnancy would occur. I simply pointed out to her that I could advise her on body needs. When life-giving enzymes are taken, the body responds.

Major problems and illnesses can afflict us when our bodies get out of balance. Another way to state this is that certain glands and tissues become too active or too inactive. A healthy network is absolutely essential to good health. The purpose of this chapter is to make you aware that many of the hormonal imbalances that men and women suffer can be arrested, or better yet, prevented. We fear those things we do not understand. Spending time educating ourselves about preventatives is time well spent.

Two big warning signals are sent out in the form of prostate swelling or painful urination. Another signal is ongoing fatigue and a feeling of being out of sync with the rest of the world. These warnings must be addressed, not ignored. This is what prevention is all about. We must learn to anticipate problems by stopping or blocking them. How do we do this? Simply by giving life back to our bodies with living foods and supplemental enzymes. Think about it. What makes the most sense? If every action, including hormonal action, is run by enzymatic activity, does it not make a great deal of sense to keep the body fortified with a continuous supply of enzymes?

Part II

Postpone Aging with Enzyme Rejuvenation

Chapter 7

Forever Young with Skin Renewal

As we grow older the body's ability to manufacture enzymes slackens. The enzyme content in the body tissues and fluids decreases with the natural aging process as well as with the onset of disease. Investigators at Michael Reese Hospital found that older people had less starch-digesting enzymes (amylase) in their saliva than younger people.

Dr. Edward Howell has stated that the faster you use up your body enzyme supply the faster you will exhaust your enzyme reserve or *enzyme potential*. "When we eat cooked, enzyme-free food, this forces the body to make all of the enzymes needed for digestion. This depletes the body's limited enzyme capacity and I believe this is one of the paramount causes of aging and early death."

Dr. Metchnikoff holds that the longevity of the Bulgarian peasants is due to the fact that their enzyme reserve was used up more slowly during their lives because their diets consisted mainly of cultured raw milk yogurt and many other raw foods. In order to avoid premature depletion of our enzyme capacity we can consume raw foods daily and supplement with digestive plant enzymes.

In other words, forcing your body to produce enzymes for digestion limits its ability to manufacture the metabolic enzymes needed to build and repair the body. Supplementing your diet with enzymes not only aids digestion, it also helps to build, repair, and rejuvenate the body, making the most of *your enzyme potential*.

We all feel pride when our skin is clear, smooth, and radiant. We cringe when it is dry, crusted, sallow, wrinkled, and lifeless. Products that offer skin renewal are like magnets for most of us. Oddly enough, enzymes are mostly ignored in this ongoing quest. However, they play a vital role in skin care and maintenance.

Enzymes are directly related to the health and appearance of our skin. They are the catalysts that work in healthy digestion, freeing the nutrients in our food that will form the building blocks for glowing skin. Enzymes are the power source behind efficient circulation that will bring nutrients to the skin. Enzymes are responsible for reducing the unsightly signs of detoxification that come through our skin. They slow the aging process that is evident in sagging and wrinkled skin. In this chapter I will cover how enzymes hold the aging process at bay. You will learn how enzyme therapy will enhance your personal skin care.

Feeding from Within

The manufacturers of commercial beauty products are well aware of the American obsession with young-looking, healthy skin. They produce creams, lotions, gels, peels, pads, ointments, and masks for a wide assortment of skin needs and problems. The prestigious brands command as much as seventy-five dollars an ounce for rejuvenating lotion. The reality, however, is this: Real rejuvenation must come from within; it is not a topical process.

The health and appearance of your skin is determined by the nourishment it receives from the blood. The skin's under layer, or dermis, is made up of collagen (protein). Collagen extracts the nutrients it needs from the blood. The outer layer of the skin, the epidermis, is not nourished from within. It continually dies and sloughs off, allowing the new skin cells of the dermis to rise to the surface. This creates an ongoing process of continuously renewing the skin that is fed from its deeper layers. That is why digestive enzymes play such a vital role in healthy skin upkeep. Enzymes determine which nutrients will be available for skin nourishment.

Real rejuvenation must come from within; it is not a topical process.

Digestive Enzymes

The skin needs fats, proteins and carbohydrates to remain healthy and vibrant. If your digestive system has an enzyme deficit, these nutrients stay locked inside the foods you eat and exit the body undigested. The skin becomes undernourished, weak and susceptible to problems.

Digestive Enzyme Deficiencies

All of us should have plenty of naturally-occurring enzymes in our diets. The digestive system makes its own enzymes to break down foods and unlock nutrients. Some of the healthy foods contain their own supply of digestive enzymes. Then why does the condition of our skin tell us that so many suffer from enzyme shortages? For one thing, food enzymes are fragile and easily destroyed. Cooking food above the temperature of 118 degrees destroys enzymes. Because the modern diet consists primarily of cooked and processed food, it is essentially devoid of food enzymes. Second, not only are many of our food choices inherently lacking in enzymes, they actually kill those already present in the digestive tract. Alcoholic beverages, sugar and caffeine destroy enzymes and leave our body's supply of them bankrupt.

Our genetic programming, which determines body type, can cause us to have difficulty with the digestion of certain foods. Supra body types cannot efficiently digest protein. Para body type has trouble with all carbohydrates, including vegetables, fruit, and starches. Estro body types do not break down fats and oils in fried and rich foods very easily. The Neuro types encounter problems with all dairy products and most carbohydrates.

When the body is lacking enzymes for any of these reasons, you will surely see it in the appearance of the skin. A shortage of protease is especially noticeable in the skin. This is because the dermis feeds primarily on protein. Protease breaks down protein to make it readily available to the cells. **Without enough protease, the skin can literally starve.**

Seeing Is Believing

A proper supply of digestive enzymes will improve the overall look and feel of your skin. The regenerative and healing power of enzymes will show up most dramatically in cases of visible skin disorders.

I want to share with you a remarkable example of healing skin with enzymes. This is my own story. As a child I had eczema all over my body. It was especially severe in the creases of my arms and legs, my eyes and mouth. At times the sores were so unsightly and painful that I had to stay home from school. Following doctor's orders, my parents would wrap me in cloths soaked in medication. I would lie in bed and cry, wearing my "mummy" prescription. My eczema was the weeping kind; if I scratched the sores or if they cracked from dryness, they would weep and then spread. I cannot tell you how uncomfortable I was or how embarrassed I felt during these years of affliction.

As I grew, eczema episodes came less often and were less severe, but I was never completely free of its dry patches and scales. Whenever I was under stress, I would break out in the corners of my eyes, mouth and the creases of my arms.

The outbreak aggravated the stress response and the cycle of discomfort, embarrassment, and cancelled appointments would repeat itself.

Later, I began to learn about the skin-healing effects of enzymes and was anxious to try them on myself. I began with a multi-enzyme supplement made up of protease, lipase, cellulase, and amylase. I took these along with my three main meals. I also took high doses of protease between meals on an empty stomach. I continue this therapy to this very day and have never once had a recurrence of the eczema that plagued me for more than forty years of my life.

Detoxify!

Along with the kidneys, the bowel, and the lungs, the skin is used by the body for detoxification. Incomplete elimination of toxins through the kidneys takes place when we have trouble with urination or simply do not drink enough water. The same problem occurs in the bowel when we suffer constipation or diarrhea. The lungs cannot do their part in detoxification if we smoke tobacco or breathe polluted air. What is left? Our skin then takes over the job. This is painfully brought to our attention in noticeable conditions like acne, eczema, psoriasis, pimples and blackheads, lyploma, and cellulite.

Our body detoxifies four ways: colon, kidney, liver, and SKIN.

Fighting Free Radicals

In addition to the natural process of elimination, the body is in a continuous detoxification battle with free radicals. These invaders roam through our body on an endless search-and-destroy mission. The pollutants we breathe, the hormones and antibiotics we ingest in our meats, the pesticides we consume on our fruits and vegetables, the bacteria, fungi, and parasites to which we are exposed are just a few of the toxins that must be discharged from our system.

There are no more powerful scavenger antioxidants than plant protease enzymes. They absorb the free radicals before they can do their damage. Any cellular debris left over after an attack is hauled away by protease molecules. These enzymes clean up free radical damage in cell membranes. They can actually enter the cell's control center and help repair devastation by free radicals to its DNA.

An ever-present supply of enzymes reduces future damage to the skin by toxins. Enzymes can even repair cells already impaired by free radicals. Enzymes are the only substances in the body that can rebuild injured cells by infusing them with oxygen and fortifying nutrients.

Signs of Damage

The effects of unchecked toxins on our skin can clearly be observed. Incomplete digestion and the adverse action of free radicals cause skin disorders of all kinds in the majority of American adults. The remedy is multi-enzyme therapy. Major enzymes work together to address skin eruptions. Any sores, blemishes, lesions, or blisters on the skin are known as eruptions. Protease, lipase and amylase combine digestive and detoxifying action to manage the following most common skin disorders.

Acne: poor digestion of fats
Eczema: poor digestion of sugars
Psoriasis: poor digestion of proteins
Pimples and
blackheads: toxicity
Lyplomas: rancid fats
Cellulite: toxins trapped in fats
Wrinkled skin
loss of skin elasticity: overall poor digestion

In addition, almost all skin hives and rashes will respond favorably to enzyme therapy.

The End of a Chronic Condition

A thirty-year-old client of mine named Nancy came to me with a moderate but pesky case of psoriasis on her hands, arms, and feet. As a nurse, Nancy needed to wash her hands often during the day which aggravated the condition and made her daily life very uncomfortable. Before calling me, Nancy had tried several topical medications, skin emollients, steroids, and even a recommended diet change. But nothing noticeably influenced the itchy, scaly skin.

I suggested the same course of enzyme therapy which I followed for my own problem with eczema. She took a multienzyme supplement with each meal and a high-dose protease between meals on an empty stomach. After three days of this therapy, Nancy could feel the positive effects of enzymes. Her energy level significantly increased. She noticed her digestive problems with bloating and belching had subsided. One week later, her skin showed marked improvement. The scales began to disappear and the color became more even. The overall appearance of her skin was vibrant and healthy. In twenty-one days the psoriasis that doctors had told her would be a lifelong condition for her had vanished. Nancy is still on the same enzyme regimen. Her skin problems no longer exist.

Youthful Skin

How youthful is your skin? Try the following simple experiment. Pinch the skin on the back of one hand between the thumb and forefinger of the other hand for five seconds. Then let it go and time how long it takes to return to normal. If it takes five seconds or less, the biological age of your skin is less than fifty. Ten to fifteen seconds indicate an age range of sixty to sixty-nine. Time exceeding fifteen seconds shows the biological age of your skin as seventy or more. It takes longer as we age. This is reflected in the wrinkles we bear from the deterioration of tissue under the surface of the skin.

Many researchers claim that aging skin is often caused by a lack of protein and a weak circulatory system. These deficiencies cause tissue, muscles, nerve fibers, and blood vessels to become rigid. They are responsible for loss of muscle tone and skin elasticity, resulting in sagging and drooping skin and muscles. This degenerative process can be slowed and sometimes even reversed with enzyme therapy.

The Protein Connection

Specialized proteins fed by protease maintain the integrity of our supporting tissue and furnish normal elasticity for the skin. The interlocking of elastin, a yellow, fibrous protein that is the basic constituent of elastic connective tissue, with collagen in the layers of the skin keeps it smooth and supple. However, as we age this interlocking system hardens. The aging process is arrested when there is an adequate supply of enzymes to feed the skin, to deliver abundant supplies of oxygen, to keep the cells plump and to allow the dermis to replenish itself with healthy new skin.

Imagine the cells and tissue of the body to look like a sponge. These cells and tissues need the moisturizing agents delivered by protease in order to function efficiently. When the body's cells lack protease, they become thirsty and they atrophy, or waste away because of insufficient nutrition. The characteristics of a sponge disappear. The cells absorb less of the necessary protein and the cellular walls break down, dry, and shrink. By this time, aging has set in.

Effective Circulation

The circulatory system as well plays a key role in aging. A sluggish system delays the delivery of needed nutrients to the skin. A blood supply low in essential nutrients sabotages the needs of the skin to stay healthy. Once again, it is enzymes to the rescue! I emphasize again. Enzymes are the digestive catalysts that make nutrients available to the blood for their journey to every cell in the body. Lipase enzymes keep the buildup of plaque and platelets off the walls of the arteries. This allows the blood to deliver nutrients and oxygen on schedule. Protease is the lifeforce of the circulatory system and its blood. Protease keeps the blood moving by cleaning up congested red blood cells that stick together in clumps. Through microscopic blood analysis we can see how, after only ten minutes, one dose of protease immediately loosens sticky cells and frees the blood flow. Because our body is composed of protein, protease feeds and fortifies the cells of the blood itself, making them stronger and better equipped to carry out the task of distributing nutrients.

Skin that is well enriched by oxygen and high quality nutrients always looks radiant, firm, and smooth.

Because the skin is the largest organ of the body, it is a primary benefactor of an enzyme-rich circulatory system. Skin that is well enriched by oxygen and high quality nutrients always looks radiant, firm, and smooth.

Young Skin at Sixty

At the age of sixty, Carl called on me because his gums were bleeding and his front tooth was loose. He had high hopes that enzyme therapy could help his periodontal problems. A high-dose regimen of protease cleared up the swollen and inflamed gums and saved the tooth. He was an avid convert to protease therapy.

Very soon Carl noticed an unexpected bonus from his daily intake of enzyme supplements—he looked twenty years younger! His skin looked smooth and soft. It lost its wrinkled and hardened appearance. Even its color was youthful. Carl phoned me one day to tell me that people were constantly telling him he looked great. Everyone wanted to know his secret. "I'm tempted to keep it a secret," Carl laughed. "But no one believes me anyway when I tell them I found the fountain of youth in enzymes."

Enzyme Therapy For The Skin

To feed, detoxify and slow the aging process of the skin, you must dedicate yourself to a lifelong program of enzyme therapy.

Foods For The Skin

The ideal way to obtain the necessary supply of nutrients is from your foods. I also advise an increase in your intake of foods rich in plant proteins. These include:

brown rice	**kasha**	**seeds**
buckwheat	**millet**	**soybeans**
bulgur	**nuts**	**wheat germ**

Raw fruit and vegetables with the enzymes in tact are good for the skin. Because protease is the enzyme most vital to healthy skin, you'll need to get it from plant enzyme supplementation. In addition, to gain the full benefits of enzyme power for the skin three other enzymes should be added to your program. Give your body lipase so it can break down fats and lipids. You will not have a healthy fatty acid balance without proper lipase balance. It also needs cellulase to break down fiber and eliminate residue from the small intestine. Add to this amylase to digest carbohydrates and reduce skin inflammations. Again, you will find these enzymes in living foods. All raw fruits, vegetables, and grains will supply your digestive system with a partial quantity of the enzymes it needs to maintain healthy skin.

Enzyme Supplementation

Unfortunately, even raw foods contain only trace amounts of protein, if any at all. That is why it is extremely difficult to eat enough raw foods each day to keep the digestive system healthy. Along with basic digestion, we want to keep a reserve of protease available for skin vitality. Therefore, I believe enzyme support is absolutely necessary for anyone who wants healthy skin. When you purchase enzyme supplements from your health food store or your own health-care practitioner, keep these tips in mind:

- **Take protease between meals to promote detoxification and purification.**
- **Take only plant enzymes**
- **Supplements must contain the three major enzymes—protease, lipase, and amylase.**
- **Always take the supplements with your meals to aid digestion.**

The Secrets of Healthy Skin

The units of activity of enzyme supplements vary from label to label. Those of you with severe skin problems may need very high units that can only be purchased from a health-care professional. In most cases, if you read labels carefully you can find a multienzyme supplement with the appropriate units of activity. These unit numbers are the minimum dosage you want for a successful program.

- **Amylase: 5500 DU**
- **Lipase: 145 LU**
- **Protease: 30,000 HUT**
- **Cellulase: 600 CU**
- **Lactase: 360 LacU**
- **Maltase: 216 DP**
- **Sucrase: 80 IAU**

Between meals to fortify and detoxify:

- **Protease: 330,000 to 420,000 HUT**

The secret to youthful skin is enzymes. This is not regarded as one of life's great mysteries. It is not complex. All of this information is well-documented, proven and simple. Your skin needs the continuous nourishment from within that only enzymes can supply.

There are so many success stories I could relate, but then this book would be too heavy for the reader to lift. Psoriasis and other skin disorders clear in short periods of time, once the enzyme therapy has started. One client had a skin disorder that is sometimes called elephantiasis. This unsightly condition is characterized by the enlargement of certain parts of the body, especially the legs and genitals. The surrounding skin hardens and becomes ulcerated. It is caused by obstruction of the lymphatic vessels, often due to infestation by parasites. This poor woman was in a great deal of pain. Two weeks after she began her enzyme therapy, she could see and feel a notable difference. Others commented to her on how much better she seemed. Working with enzymes is so exciting. I feel truly blessed to be a part of this wonderful healing art. It makes so much sense to me, because it works!

Summary: Enzymes For The Skin

Primary Enzyme:

Protease: **Breaks down protein foods that feeds the cells of the dermis; also improves distribution of all nutrients to the skin.**

Additional Enzymes:

Lipase:	keeps skin cells plump to reduce wrinkling
Amylase:	reduces skin inflammation
Cellulase:	breaks down fiber and allows nutrients access to the skin.

Protease, lipase, amylase, and cellulase are found in living foods.

Chapter 8

To a Longer, Healthier and More Youthful Life

Longevity Potential

Aging can be delayed. Tests conducted by Dr. Howell and his colleagues clearly demonstrate that aging and debilitated metabolic enzyme activity are synonymous. Other very exciting research continues to shed new light on how our bodies rid themselves of damaged and aging cells. This offers renewed hope that someday we will be able to manipulate these processes in order to further delay aging. In the meantime, we already have some fascinating information about the role enzymes play.

We know that the major cause of aging is cellular deterioration due to excess toxins in the body. This, in turn, reflects a weakened immune system from the destruction of free radicals causing a general contamination of the blood. I find it very interesting that nearly every problem or adverse condition hinges on virtually the same deficiencies. In the previous chapter you learned how current studies on plant enzyme supplementation cause the skin to look more youthful.

I ask my elderly clients to adopt an antiaging stance, then attack with gusto! It is simply not enough to eat well and exercise. Everyone has to add nutrients to his or her diet, preferably in herbal form or raw, living foods. Everyone has to get enough sleep. Everyone has to take time each day to meditate or spend time in a relaxed state. Everyone has to eat those foods that are appropriate for his or her genetic body type. Finally, everyone has to take plant enzymes with meals to digest food and between meals to repair the system. Those who faithfully remain on their plans tell me they have never felt better or more energized since they were kids. That is because, first and foremost, a good program of enzyme supplements helps your body get the most out of the foods you eat. By fortifying and revitalizing your organs and tissues, you achieve balanced health and antiaging support. Enzymes assist the body to heal itself.

Why Enzymes Keep the Body Feeling Younger

After reviewing results from a variety of tests on aging, Dr. Howell, who began his study of enzymes in 1930, concluded, "Humans eating an enzymeless diet use up a tremendous amount of their enzyme potential in lavish secretions of the pancreas and other digestive organs. The result is a shortened life span (sixty-five years or less as compared with one hundred or more), illness, and lowered resistance to stresses of all types, psychological and environmental. By eating foods with their enzymes intact and by supplementing cooked foods with enzyme capsules, we can stop abnormal and pathological aging processes." My favorite quotation from the work of Dr. Howell is, "If you take in enzyme reinforcements during the younger years, your enzymes at eighty years will be more like those at forty years."

Aging has a great deal to do with decreased enzyme reserves. Enzyme therapy slows down the aging process by building up the enzyme reserves and quenching free radicals. Anytime we suffer from an acute or chronic illness, it is almost certain an enzyme depletion exists. Those with digestive disturbances, endocrine gland imbalances, blood sugar imbalances, diabetes, obesity, high cholesterol, high triglycerides, and stress-related problems will need supplemental enzymes. Enzymes aid in detoxifying. They free up more metabolic enzyme energy to enhance detoxification. If you have a very busy life, fly or travel a great deal, and cannot eat those foods common to the area where you live, your enzyme reserves are very depleted.

Anytime we suffer from an acute or chronic illness, it is almost certain an enzyme depletion exists.

How a lack of enzymes speeds the aging process

Medicine has discovered little about how to retard the inexorable process of aging. Biologists do not even agree on its definition. We do know that cells tend to die off faster than they can be replaced. After the age of thirty, the body's water content and energy reserves noticeably begin to decline. Bodily efficiency decreases roughly one percent per year. Tissues waste away, enzymes disappear, mutations damage the genes and organs wear out, one after another. Typically, in the body of an eighty-year-old man, fifty million cells are dying each second, while perhaps only thirty million new ones replace them. His muscles have lost 30 percent of their former weight, his brain has shrivelled by 10 percent, and nerve trunks have shed 25 percent of their fibers. Each breath takes in 50 percent less oxygen, each heartbeat pumps 35 percent less blood, the blood absorbs oxygen 60 percent more slowly, and the kidneys, loyal team members that they are, have sacrificed their efficiency by half to help other organs worse off than themselves.

That our bodies continue to function in spite of all this speaks wonders for this biological miracle of one hundred organs, two hundred bones, six hundred muscles and trillions of cells! To carry on a detailed discussion of the way all these parts work would bewilder me and the reader. It is sufficient to know that none of these parts or processes can possibly work without enzymes. Enzymes are the energy and substance that make life possible. They are a necessary component of every chemical action that takes place in the human body.

Invariably any discussion about enzymes results in the question: Do we loose our ability to produce enzymes because we age or do we age because our ability to sustain production of enzymes decreases?

Part III

The Enzyme Effect on Body Shape and Weight

Chapter 9

What Is Your Body Type?

What Works for One Does Not Work for All

Have you noticed how people seem to be one of four basic shapes? At any shopping mall in the Midwest, for example, it is not unusual to observe many shoppers who carry most of their weight in their tummy or upper torso, but not their buttocks. A cowboy who has trouble holding up his jeans because his tummy is bigger than his hips falls into this category. You will also see women with large breasts and shoulders but small, flat buttocks. On the West Coast, health-conscious men and women tend to carry their weight evenly all over their bodies. Rather than suffering from weight problems, they tend to have trouble with fatigue. In the East, many people have large hips wider than the shoulders, and oversized thighs in women. Everywhere you will find those whose body shapes mature later. This type has a boyish or childlike shape. This does not mean that these shapes are peculiar to specific geographic locations. However, when a particular nationality is concentrated in one state, this often accounts for a preponderance of one of the body types in that locality. We inherit our body shapes with their strengths and weaknesses from our ancestors.

What does this have to do with you, or this book? Everything, because with different body shapes come different needs and different enzyme potentials. The key word is different. Because of this difference, we cannot have one perfect diet that works for everyone.

Now that you are thinking about these varied body shapes, consider the distinct cravings that go along with each of them. One man may be drawn to very sweet foods, while another strongly dislikes desserts. He much prefers a good steak rather than a slice of chocolate fudge cake. Doctors ask their patients about their own health and that of their families. These doctors are very aware that we inherit certain weaknesses and disorders. Some medications and foods affect people very differently.

Dating back to the ancient Egyptians, over thirty-four books have been written on the subject of body typing. Ayurvedic medicine, originating in India five thousand years ago, is completely based on body typing. The study, or science of body typing is not new. Health professionals have always been very aware of the differences in body type and muscle tone, and their similarities. I believe that one of the major reasons any body type suffers weaknesses and becomes out of balance is due to enzyme deficiencies. These are inherited, and compounded through poor eating habits and stress on the system.

How Your Body Shape Affects You and Your Loved Ones

All four of the basic body types have certain weaknesses common to all of them. You or any other members of your family may carry the greater part of your weight in your upper torso and tummy. Your buttocks are flat and your legs are strong. Quite probably, you are a family that enjoys meat as the main part of your meals. You and your family could very well suffer from lower back problems, acidic stomach, pain in the left shoulder, constipation, stress, and high blood pressure.

For those who have a family tendency toward weight that is evenly distributed all over, you will find hypoglycemia, low blood pressure, ulcers, allergies, colon problems, anxiety, and fatigue to be common ailments. You gravitate toward sweets, breads, pasta, and chocolate. If you gain weight, it will invariably be carried in your abdomen area.

The family that enjoys tasty, strong flavored foods like ethnic foods, wine, ice cream, heavily spiced food, or lots of salt and pepper has its own set of disorders. Any or all of you may suffer from kidney problems, female problems, prostate, skin irritations, sinus and mucous trouble, cysts, gallbladder distress, or indigestion that might include constipation to diarrhea.

I cannot leave out one of the most interesting body types. This is the one in which the family has serious difficulties with lactose intolerance. The cause of this is, of course, serious cravings for all dairy products. Some of the common maladies of this type are milk allergies, spastic colon, other allergies that range from foods to airborne microorganisms, aching joints—especially the knees and a continual feeling of exhaustion without knowing why.

Quiz: What Body Type Are You?

In these descriptions, do you recognize yourself or someone you know? Are any of these families yours? Do you feel you belong in each one of these categories? If you answered "yes" to any of these questions, and you will, it is because we inherit our body types from our parents. If your parents married outside of their own nationality, you will probably have a combined or overlapping body type. For instance, an Irish man and an Italian woman get married and have a family. We know that the Irish are evenly shaped and love sugar, potatoes, and alcohol. Italians, on the other hand, are known for their curves. They indulge in rich, strong-flavored, and highly seasoned foods. The child will always have some of the strengths and weaknesses of both parents. The Italian mother will cook foods she likes and see to it that her husband has an ample supply of breads and desserts. This may all sound delicious to you—it certainly does to me! However, as an adult, a child from this marriage has developed a weak immune system due to the stressful job he has. He seeks the help of a nutritionist.

What are his best food choices? Should he be a vegetarian? These are the exact questions I had to answer for myself when I became a health-care professional. I realized very early that what worked for one does not necessarily work for another. There is a very good reason for this. We are not all the same body type, nor do we have the same needs and strengths.

The Four Major Body Types

I will refer to the four body types by number. It is not that one is more important than another. Now that I have worked with body typing for over ten years, I find it is best to begin with the category in which I see the greatest number of body types.

Body Type One

In conventional medical practice, the body types are referred to by the name of the gland that rules each one. Type One is known as Para, for the parathyroid or thyroid gland.

If six or more of the twelve characteristics listed apply to you, this is your category or body type. It is not a good idea to eat the foods listed, because they are the source of many of your health problems. However, you continue to crave them.

1. **You gain weight evenly, from the top of your head to your toes.**
2. **If overweight, your excess fat is in the stomach and waist area. Men have "spare tire," women hold fat in their buttocks.**
3. **Your shoulders and hips are the same width, and in proportion to each other.**

4. You crave sugar in the form of desserts, breads, pastas, fruits, and vegetables.
5. You are drawn to chocolate and coffee.
6. You are fatigued by midday and sometimes get up tired.
7. You suffer depression or mood swings.
8. You often have cold hands or feet.
9. If female, you have PMS.
10. Your blood pressure is low.
11. You have hypoglycemia.
12. Allergies are common for you.

Body Type Two

The controlling glands for type twos are the gonadal system. Type Two is known as the Estro body type. The following list of traits defines this body type. Again, if six or more of them apply to you, this is your category.

1. Your shoulders are narrower than your hips.
2. In women, the greater part of the weight is in the buttocks and upper thighs.
3. In men, your excess weight is in the abdominal region.
4. Your favorite foods are tasty, or strong-flavored. These include spiced, creamed, salty, peppered, or smoked foods.
5. You drink wine or a beverage other than water with your meals.
6. Your favorite desserts are rich, like ice cream or chocolate.
7. You enjoy fried fish, chicken, onion rings, and French fries.
8. You like Italian, Oriental, and Mexican cuisines.
9. You suffer from indigestion.
10. Gallbladder problems are common.
11. Your bothered by kidney and bladder infections.
12. You have a tendency toward cysts or skin disorders.

Body Type Three

The third type is known as Supra, for the suprarenal or adrenal glands. The same criteria apply for this type.

1. You carry your weight in your tummy and upper torso, including the back.
2. Your buttocks are flat and do not have any shape.
3. Your legs are strong and firm at any age.
4. A meal is not complete for you without meat or some kind of protein.
5. You salt your food without tasting it first.
6. You gravitate to all proteins: Lunch meats, sausage, fish, poultry, nuts, cheese, and eggs.
7. You suffer from stiff joints or pain in your left shoulder.

8. **You must drink at least one beverage with your meal.**
9. **You have hypertension, or high blood pressure.**
10. **You feel stressed, rather than fatigued.**
11. **Constipation is a big problem for you.**
12. **Gas or flatulence is a way of life.**

Body Type Four

The fourth type is known as Neuro. It gets its name from the master gland of the endocrine system and the pituitary gland. These glands control the hormone production of other glands and indirectly affect the neurological system. If you can answer yes to four or more of these questions then this is your body type.

1. **You do not feel that a meal is complete without dairy products.**
2. **You are drawn to starchy foods and sweets.**
3. **You enjoy dairy products but recognize a milk intolerance.**
4. **You have the same body size as when you were a teen. Your shape may be described as boyish, wiry, or having baby fat, but it has remained the same into maturity.**
5. **You matured later in life or are the youngest looking member in your family.**
6. **You have an on-going bowel problem (constipation to diarrhea) including spastic colon.**
7. **You have soft fat all over the body, not just held in one area.**

Many people see themselves in more than one category. However, all of us are more like one than any of the others. This is known as our primary body type.

What Your Body Type Tells You About Health and Diet

You may have already concluded that the foods you crave are the very ones that create most of your problems. The reason for this makes a great deal of sense. We all want to eat foods we enjoy; foods we were given as children. But as we grow older, our ability to produce the necessary enzymes to digest our favorites decreases. Many of us are born with certain enzyme deficiencies. Poor eating habits, developed in childhood, along with stress only compound the problem. My experience, and those of my clients, shows that when depleted enzymes are restored to the body, it comes back into balance. A balanced body heals itself.

When a client enters my office, my staff has become very proficient at recognizing what enzymes are lacking just from observing the client's body type. Furthermore, the staff can correctly anticipate how the client will fill out the questionnaire. You do not have to have Ph.D. written after your name a half dozen times to see the link between enzymes and body types. We all need plenty of naturally occurring enzymes for proper digestion. The digestive system creates its own enzymes to break down foods and unlock access to their nutrients. Some of the foods we eat contain these digestive enzymes. Then why do we suffer from enzyme shortage?

There are a variety of answers to this important question. Food enzymes are fragile and easily destroyed. If we cook our food at a temperature higher than 118 degrees, many enzymes are sacrificed. Most of us eat only cooked or processed foods, both of which are devoid of enzymes. Some of our food choices actually kill enzymes that normally thrive in the digestive tract. Alcoholic beverages, sugar, and caffeine all destroy enzymes. Our body type determines which foods will create digestive difficulties. It is good for us to take a closer look at the genetic programming that dictates our shape, size, and food cravings. Each person has a genetically programmed ideal weight. If we are overweight we remember when we felt and looked our best.

Body Type One craves sugars and those foods that form glucose (sugar) in the system. It is easy for you to want to be a vegetarian if you are this body type, because you will only desire starch and sugar. Starches are those carbohydrates derived from grains. Some of these are rice, wheat, potatoes, pasta, bread, and cereals. Vegetables and fruits are carbohydrates as well, and break down into sugar or glucose. Most foods contain trace amounts of carbohydrates (sugars). These include granulated sugar, maple sugar, honey, and molasses. Simple sugar is fructose or fruit sugar. Double sugars are sugar cane, beet sugar, maltose (malt sugar), and lactose (milk sugar). All ripe fruits contain natural sugar. Enzymes critical to the breakdown of these foods are amylase, lactase, sucrase and maltase. Amylase is the most important one because of its ability to break down starch. Body Type One is deficient in amylase.

Food cravings create a vicious circle. The more exhausted a Type One feels, the more he or she will crave carbohydrates and starches. He must have his "fix" of candy, especially chocolate. Chocolate has a huge following of addicts because it is a blend of over five hundred flavors. It stimulates the taste buds more than almost any other food. Its melting point is just below the temperature of the human body, so the texture and sensation of it melting in our mouths have great appeal. Chocolate is a perfect combination of 50 percent fat and 50 percent sugar so it can stimulate certain brain chemicals with perfect equanimity. One of these is serotonin, a chemical which fosters calmness and mood stability. The other, of course, is those famous endorphins; those chemicals that send our "feel good" messages to the brain. Endorphins give us a high level of energy and a feeling of euphoria. When our serotonin level is depressed, we crave potatoes, pasta, bread, and other sugar foods. Body Type One has been found to be low in serotonin. This absence is hereditary.

I do not recommend to any of my clients that they go cold turkey and stop eating their problem foods overnight. First, it is so important to learn why these cravings exist. The next steps are to control them so the harm they are doing ceases. This is where the enzyme amylase works wonders. If you make the choice to eat your toxic foods, then be prepared to supply your body with the enzymes that will neutralize them. A Body Type One will always find a way to get a "fix" of the food he or she craves. Now, with supplemental enzymes there is a way to control these cravings. Carbohydrates and starch foods that are improperly digested cannot be used to feed the brain. Instead, they ferment in the system. This creates everything from inflammation to fatigue. This body type has a susceptibility for colon cancer, hypoglycemia, and nervous disorders such as depression and anxiety.

Body Type Two tends to be attracted to foods that are strongly flavored, like spicy, smoked, or salted foods. These foods stimulate the gonadal system (the sexual organs), which creates a feeling of control and high energy. As with other imbalances, these feelings only last for a short while. The craved foods for this type wreak havoc with the gallbladder. I do not need to spend very much time on fat facts. The media has made our entire nation very aware of the harm and damage that fat does. Its victims, time and time again, have serious heart problems due to plaque in the arteries, obesity, and kidney and gallbladder stones. The human body needs only about one tablespoon of fat each day, although the Body Type Two consumes eight times that amount. Some of the other body types are in danger from not consuming enough fat, which leads to skin ailments. However, this does not apply to Type Twos. The skin disorders they suffer are caused by other conditions. Attracted to chocolate as well as Type Ones, they eat it for the same reasons. Like Type Threes, they get a great deal of pleasure from eating the marbled fat in meat.

The Type Two immune system is compromised by the characteristic bloating, water retention, and constipation. These body types produce a lot of mucus, and suffer more constipation than any of the others. Because they often go from constipation to diarrhea, they do not even realize they are constipated. These are the body types who will come to my office and laugh when I ask how many bowel movements they have in a twenty-four hour period. They truly have no comprehension of what I am trying to say. I explain to them that a healthy system will have a bowel movement following each meal, and a minimum of two per day. I find that a Type Twos may go for a whole week before there is any movement at all. This can be followed by several days of nausea and diarrhea. When the diarrhea stops, several days more of constipation may elapse. These sufferers are in dire need of lipase. Lipase breaks down fat during digestion and when taken between meals it cleanses the system. If fats are not digested properly, they turn rancid in the intestinal tract. This body type is prone to diabetes, hormonal problems, and skin conditions.

Body Type Three prefers protein to other foods. Like the others, this type enjoys some chocolate. However, Type Threes get most of their chemical cravings satisfied with protein. Type Threes tend to have greater muscle mass. Excess protein creates problems with the kidneys and liver. If these two vital organs are not working properly, we will die. Our system has only four detoxification centers, so it behooves us to take good care of them. One of these is the colon. Remember that if protein is left undigested in the intestines, it putrefies. The other three are the lungs, the skin, and the kidneys, the primary function of which is to rid the body of excess protein. I have concluded that psoriasis is a condition caused by poor protein and fat digestion. Type Threes require less protein than any of the other types because of their muscular structure and a poor ability to break down protein. Through many years of experience, I have discovered that when any body type ingests the enzyme in which he or she is the most deficient, almost immediate relief from bloating and water retention occurs. Protease, the protein enzyme, is the one most lacking in Body Type Three. This body type has a tendency to contract hypertension, heart disease, constipation, and lower back problems.

Body Type Four is given the food most craved right from birth. These are the people who gravitate toward dairy products. If they were fortunate enough to be breast fed by their mothers, their potential for handling other foods is greatly improved. Most Type Four people with health problems were not fed their mothers' milk. Often, the mothers were unable to provide enough milk. This seems to be a sad situation, but more than likely the mothers themselves were Type Four and suffered extreme food allergies. Body Type Fours will drink milk until told they are allergic or lactose intolerant. Extreme lactase deficiency characterizes this group. In addition, they need lipase to break down the fats found in milk products. I have seen firsthand the frustration of this body type. They go to many different doctors in vain attempts to find out why they are sick or exhausted all the time, but they never find satisfactory answers. A tendency to mature more slowly than other types keeps Type Four high school boys from being developed enough to be successful in sports. The girls will also not reach puberty as quickly as the others. Later in life they look younger than their peers and keep the body shape they had in high school. If they gain weight while young, their soft flesh is more like baby fat. They almost always have similar bodies as adults. The other extreme is a boyish, wiry body in both males and females.

Type Fours have trouble digesting any fiber or dairy products. Their choices are limited because meat generally has no appeal. These body types can have slight to very severe colon problems because of an inability to digest most foods. To counteract the stress put on the colon, they are encouraged to eat more fiber. This makes matters worse. Fiber is made up of cellulose. Yet the human body is incapable of producing its own cellulase, an enzyme that breaks down cellulose. When foods containing cellulase are processed, the cellulase so desperately needed to break down the fiber is destroyed. Another vicious circle is created. Luckily, if given the proper formulation of enzymes, digestion becomes much more efficient. The skin begins to clear and the hair glows with a healthy sheen.

When we feel ill, most of us seek the services of an allopathic physician. Allopathy, as opposed to naturopathy, is a method of medical practice that treats symptoms. Within this framework, a patient is treated with remedies that produce the opposite effect of those created by a disease. Under the care of these doctors, who are responsible medical personnel, we often feel better for a time. However, in many instances, the symptoms return again and again. Unfortunately most medicines are enzyme inhibitors which force our body to work harder than necessary to fight off infections and diseases.

When we feel ill, most of us seek the services of an allopathic physician. Allopathy, as opposed to naturopathy, is a method of medical practice that treats symptoms.

Unsatisfied, we turn our attention toward the alternative health care community, exemplified by health food stores, nutritionists, and dieticians. By now we may be experiencing rapid weight gain or loss, constant lethargy, or exhaustion; a mountain of debt from all our expenses towers over us, and we become angry. I am sure at least part of the following nightmare has something to offer that relates to your own personal experience.

Have you experienced this? Someone has just told you that eggs are bad for you. You blink, because you had scrambled eggs and hash browns cooked in grease, along with fried bacon or sausage, served every single Sunday morning while you were growing up. No; now it's okay to eat eggs again, after you've struggled to give them up. They have **good cholesterol** as it turns out. Protein is important; no, don't eat protein! Eat according to the food pyramid and you'll be fine. That's all wrong, but here, I have a new pyramid for you to try. Isn't it great, now we have genetically altered foods, the wave of the future, say the scientists. But are they dangerous, you ask? I don't know, the government is not required by law to tell us whether a food item has been genetically altered. The top soil blew away years ago—but be sure to eat agricultural products. Never eat dairy, it causes too many allergies. Fast and you will cure all your ailments. Don't fast, your body will begin to consume itself. Eat margarine instead of butter. Now we have conclusive proof synthetic fats are much worse than natural ones. Be sure to read those labels in the supermarket. Never mind that the small amounts of toxins added to our food do not have to be reported. Are you confused and frustrated and angry yet?

In Chapter Ten you will find out how knowledge of your body type can lead to a state of *optimum health.* I will answer many questions about this critically important topic. You will understand why you hold fat where you do . . . what foods give your body type energy and what foods create fatigue . . . why we all have food cravings . . . why you react differently from another body type to the same food item . . . how your emotions are affected by body type . . . the best time for you to eat . . . why you gain weight . . . what enzymes you need . . . why you feel brain fatigue or anxiety . . . why you suffer nausea from certain foods . . . and a host of other points worthy of consideration.

Chapter 10

Shedding Excess Fat and Weight . . . and Enjoying It!

One in three Americans is overweight, tipping the scales at 20 percent more than his or her ideal weight. One in four of us is downright obese; defined as at least 30 percent heavier than our healthy weight. Those of us who carry excess weight desire nothing more than finding a way to lose fat without changing our lifestyles. We do not want to go on diets, or change our outlooks, or exercise or control our appetites. I state these things with the voice of personal experience. I have read every newspaper article, I have perused the bewildering array of books on weight loss at my local bookstore, I have bought every self-help pamphlet by the cash registers in supermarkets, and I have discussed the problems of weight gain and loss with countless overweight clients. It is nearly impossible to pick up a magazine and not find at least one article on weight loss featured in it. We as a nation spend over thirty-five billion dollars a year on weight loss programs, and the only thing we are really loosing is *ground.*

Recently I read about the latest weight loss craze that swept through Tokyo. People were rushing to buy a Chinese-made soap that supposedly washes excess weight away! Officials had already seized more than ten thousand bars of the so-called "Soft" soap from travelers who had exceeded the legal limit of twenty-four bars per person. The soap costs twenty dollars for a 5-ounce bar. The pharmaceutical company that manufactured it exported one and one half million bars in less than eleven months.

Another article published in the local Houston paper was headlined "Drug That Melts Fat Draws Crowd" (Houston Chronicle, 1995) Here is a quotation from that issue. "How much would you pay for a drug that melted fat, calmed a savage appetite and replaced dimpled sag with youthful lean? What is it worth to shrink from a size eighteen to a six, from a fifty-two inch belt to a svelte thirty-four, without ever breaking a sweat? The answer: A great deal. A very great deal. Experts in the field say that in America, the sky is the limit.

People deplore fat bodies, particularly their own. Thin is in." This latest miracle drug required the user to receive daily injections. The paper was besieged with so many calls they had to add more phone lines.

My favorite radio station plays an ad in the morning about a pill that promises to curb the appetite and burn fat. I phoned for more information and finally got through to an operator. She told me many others like her were trying to take care of the phenomenal number of people tying up the lines. She explained that every time they run the ad, people clamor for appointments to obtain the pill. It is a simple, common herb, and yet it sells like crazy. The big craze was, Fen-Phen, marketed by medical doctors.

Almost every day there was an article in the paper about the negative side effects of Fen-Phen. Yet there were ads in the paper and billboards along the highway with doctors willing to prescribe it.

A major women's magazine featured an article on parasites. The following statement was taken from it. "Do you ever wish there was a better way to get rid of your fat—something less strenuous than exercise and more elegant than liposuction? Well, hold onto your saddlebags—scientists are trying to genetically engineer a parasite that will feast on fat as it passes through the intestines. Presumably you swallow the baby beasties in capsule form and they munch as they mature. No news yet on what happens if the little critters decide to expand their diet." All these examples sum up the sad story of how we feel about carrying excess pounds.

I believe the saddest thing I heard was on a national television show. In interviews with children, the questions concerned weight gain and loss. One of them was, "If you were given a choice to lose one of your hands or be fat—what would you choose?" All the children interviewed answered that they would rather give up one of their hands than be fat.

What is more, another question asked was, "Would you rather be friends with a nice, sweet, honest, and loyal person who is fat, or someone who's mean, treacherous, dishonest, and cruel but who is very thin?" Again, they unanimously chose the thin person as the preferable friend!

Nearly every one of my clients who seeks help with weight loss suffers from inadequate metabolism. Worse yet, the excess weight causes an even lower metabolic rate. Why is your metabolic level so important? Technically, metabolism is defined as the sum total of all changes of energy from one type to another in your body. These changes can be from electrical to chemical energy, from chemical to physical or any one of a number of others. This simply means that metabolism takes place when fats and waste material are broken down, the nutrients are assimilated and new tissue and cells are created. Like the furnace in your home, your metabolism regulates the heat that is converted to energy in your body.

Depending on the food you eat, you can burn more energy and fat by "turning up" your rate of metabolism. The more heat there is, the more calories you burn for energy. It is essential to have a good strong metabolic rate in order to rid your body of unwanted fat. I have already discussed the ways in which a steady program of enzyme supplements can restore your metabolism to the optimum level for the job it is required to do. Boosting the metabolism should not be confused with those treatments that prescribe stimulating the system. Some of those methods work for a short time. What happens when the stimulus stops? Weight gain, of course. There are four other steps you must take to successfully lose weight, in addition to safeguarding your metabolism. (1) Exercise. (2) Strengthen your immune system. (3) Control your appetite. (4) Change the foods you eat. Now I will explore the reasons why.

Metabolism is defined as the sum total of all changes of energy from one type to another in your body.

Avoid Typical Diet Regimens

How many people do you know who have become yo-yo dieters? Up, down, up, down, pounds on, pounds off. Several studies confirm that roller coaster dieting makes it more difficult to lose weight as the on again, off again syndrome gains a stronger position in the hapless dieter's life. With each prior weight loss and gain, it is harder and it takes longer to lose weight. When pounds begin to drop, the body automatically shifts into a self-protective mode. The endocrine and nervous systems call out for reinforcements. Through extremely intricate sensor and response mechanisms, these systems interpret the new weight loss as starvation. Working together, they reduce the metabolic rate so your body cannot use up as many fats as usual. Unfortunately, when you start eating normally again, your metabolism remains in a depressed state. The frustrating outcome of this story is the more you diet, the lower your metabolic activity. Hence, it becomes much more difficult to lose weight and much easier to gain it. What a disaster!

I have observed the body shapes and types of those who write books and articles on weight loss. Naturally, we spread the word about what works for us. **Body Type One** authors will tell you to stop eating bread, desserts, and limit carbohydrates. They discovered that when they went on well-known high protein diets they lost weight. Therefore, they will continue to eat large amounts of protein and eliminate carbohydrates.

Body Type Two writers will educate you about fats and how the body first burns up carbohydrates. Then, it turns them into fats. Carbohydrate shortage can be regulated, but fat storage cannot. This body type advises you to cut back on fats and then you will lose weight. High fiber diets sometimes work well for this body type.

The Type Threes have lost weight because they became vegetarians. This works beautifully because they feel better as well. Consequently, they will tell everyone to become a vegetarian.

Body Type Four authors find the high protein diet works well for a time. Then, the immune system begins to fail, and they become ill with respiratory problems. These authors will write about glandulars (natural derivatives from glandular secretions) and antioxidants (to neutralize free radicals) because they have proved to be effective treatments for them. They cannot get very excited about diet. Rather, their focus is on whatever treatment they can find to fend off toxins.

One time, when my practice was located in Denver, another new fad diet hit the newsstands and TV ads. Six weeks later, there was a notable increase in the number of new clients calling to schedule appointments as quickly as I could see them. After recording their health and diet history, I was intrigued to find that almost 90 percent of them had immediately started to follow the new diet plan. Now, they were all very fatigued and discouraged because the results were not what was promised to them. All of them seemed to know someone else who had been on the diet and who had done beautifully. Therefore, they could not understand why they did not have the same results.

Fad diets are fads. If they worked for everyone, we would stay on them indefinitely and would not have to look anywhere else for a suitable program.

Remember, what works for one does not work for all. The fad diets are fads. If they were simple to follow, or if they worked for everyone, we would stay on them indefinitely and would not have to look anywhere else for a suitable program. When you hear someone say that fat people get fat because they eat too much, that is not always true. Many are deficient in digestive enzymes so they cannot properly digest food or burn fat. It is my firm belief that without a regimen of enzymes the dietary choices we make are to no avail. None of what we try to accomplish will last, because the basic ingredient is not there. Enzymes are that "missing link." We human beings need enzymes working efficiently throughout our bodies all the time!

When **Body Type Ones** become vegetarian they feel great. At last, they say, we have reached Nirvana. They are not very selective in what they eat in the beginning, but rather they enjoy all fruits, vegetables, and grains. After a while, they notice that they are loosing hair, their nails are splitting and cracking, they cannot shake a feeling of fatigue, and their skin may be breaking out. Depression sets in, along with PMS in women, mood swings, and inflammation. You have met these people; they have convinced themselves they are practicing good health, but their appearances tell you otherwise.

Can a Body Type One be a successful vegetarian? I do not believe so. Careful blending of grains and soy or other protein derived from legumes is essential to obtain vital amino acids. This all takes time and effort, and a very firm commitment. Only rarely will you find Type Ones with enough discipline to do this. They may achieve success with vegetarianism. In most cases, however, this kind of diet will compromise the immune system for Body Type One. The need for protein in this body type is greater that in type two and three. If it is not satisfied, the Type One will develop a protease deficiency caused by not getting enough protein. The body will be forced to use its own protein, a process known as catabolism. This can seriously undermine and weaken the entire system if it is left unchecked. If you are a vegetarian because of your spiritual beliefs, it is important for you to learn how to correctly blend vegetable proteins in order to meet your needs. Please remember that it is your inherited genes that require protein foods for this Body Type. To build the immune system for Body Type One, take a multiple plant enzyme formulation containing protease, amylase, lipase, cellulase, lactase, maltase, and sucrase. Between meals, use a high potency protease product to purge the toxins from your system.

Always keep in mind that food proteins are of great importance for good nutrition. The body does not store protein. Mixed proteins need to be eaten within two hours of each other to be used by the body as complete amino acids. These are the proteins that are used as building blocks. They help build and repair tissue, particularly in the muscles and organs like the heart, liver, and kidneys. The primary sources of protein are meat, eggs, fish, and most dairy products. The lesser or incomplete sources are soy beans, wheat germ, nuts, seeds, brewer's yeast, grains, and legumes (beans and peas). These foods need to be mixed and eaten at the same time in order to provide complete amino acids (*i.e.,* black beans and rice).

Body Type Twos are always lipase deficient. If they are overweight, they will be deficient in other enzymes along with the other three body types. Improper eating habits and stress can cause these deficiencies. A formulation high in lipase activity will help rebuild all the body's systems. If it is fortified with the other enzymes it will enhance the endocrine and nervous systems as well. Body Type Twos must take high dosages of lipase to convert their fat into energy. Those of you who are this type should regard "low or no-fat foods" as a big, red flag. They are loaded with sugar!

Body Type Three suffers depleted protease for its major deficiency. As with the other types, you will have an overall enzyme deficiency if you are overweight. Your tendency is to eat large amounts of protein. If you continue to do so, you need a high dosage of protease between meals as well as with them. This body type eats foods of all categories. Therefore, I suggest a good multiple plant enzyme capsule every time you eat something. Your characteristic aches and overall stiffness will lessen with the use of plant enzymes.This is the genetic body type that does beautifully as a vegetarian.

Body Type Four has deficiencies in more than one enzyme and many times all the known ones. Earlier, I state how important lactase, lipase, and cellulase are for this body type. The efficient use of these can be strengthened by taking protease with them. This is the most complex of all four body types.

If there is one thing all the weight loss experts agree upon, it is that any program's success can only be assured if it is ongoing. For example, you cannot take enzyme support for a few days and expect to rebuild and heal all the abuse inflicted on your body over a period of many years. The way we were raised by our parents, what kind of body we inherited from them, how we were treated, what kind of diet we were fed . . . these are all critical factors in the state of our health today. But they do not change our genetics. We use up enzymes continually, so the supply must always be replenished. Enzymes differ from vitamins and minerals in that no nutrients from any substance we ingest will ever accomplish their intended purpose without enzymatic action. Our need for enzymes should never be regarded as temporary. They are necessary for our good health and well being from the moment we enter the world until death.

Identify and Strengthen the Metabolism Unique to Your Body Type

During the metabolic process of **Body Type One** a great deal of insulin is produced. When you eat fatty food with sugar in it, the sugar accelerates the absorption of fat by forcing more insulin into the bloodstream. The insulin is now transporting fat and sugar into the cells. This explains weight gain from breads slathered with butter and jam, iced doughnuts, pasta with sauces, desserts, and sugarcoated cereals. All these combined insure that Type One will keep plenty of fat on the body.

The **Body Type Two** burns fat slowly. This is a genetic trait. Slow fat burners do not produce as many of the enzymes needed to convert fat to energy. Burning calories requires the presence of oxygen. More oxygen is needed to burn fat than to burn carbohydrates. This body type may have so much fatty tissue that it is more difficult to deliver fat-burning oxygen to the blood vessels leading into these fat-laden cells. The spices and salts consumed by Body Type Two make it even more of a challenge to burn fat.

Body Type Threes have muscular structures, particularly in their youth. As they age, even if they become overweight they will not have any cellulite in their thighs or buttocks. This body type desires protein or alcohol to stimulate them and help them feel they are in control. Each time they lose weight, it is gained back more quickly. This is now a proved fact. Through both human and animal studies, researchers have shown that weight regained after a low calorie diet comes back as fat. If you are on a lean muscle tissue diet, you will lose lean muscle tissue. The body reacts by pulling in its energy reserves as quickly as possible. The re-acquired weight shows up as fat in the abdominal area. This is a more unhealthy place to accumulate fat than the "saddlebags" that characterize the Type Two weight gain. A concentration of abdominal fat begins to interfere with insulin regulation, thus increasing the risk of heart disease or diabetes.

Body Type Four is the most complex of the body types. This type is at a disadvantage from the very beginning. If you are a Type Four, you will mature later than the others, because of a slight deficiency in hormonal or endocrine support. You are a mucus maker which means more allergies plague you. Prone to respiratory problems along with the allergies, you burn fat and calories much slower than the other three types. Some obese children who cannot lose weight because of a chemical imbalance belong in this category. However, the body Type Four is, for the most part, unable to gain weight; you manage to maintain a good, healthy weight with little effort. You are in that small minority who looks for ways to gain weight and increase your energy levels.

How To Control Your Appetite And Cravings With Enzymes

Your endocrine and nervous systems work closely with one another to regulate your appetite. When the endocrines sense that the body has had enough food, they stop the cravings for more. They also know precisely the correct caloric intake for you. However, if you ingest food with a high calorie content but lower levels of nutrients and enzymes than your body was expecting, your endocrines automatically respond by over-stimulation of the digestive organs. These organs interpret this stimulation as a demand for more food, which is false. The increase in hormonal secretions causes overeating, obesity, and exhaustion.

One way to control your appetite is to take the enzymes required to digest the type of food you crave, resulting in the subsidence of the craving. For example, if you crave sweets, taking enough of the enzyme amylase will stop the craving.

There are other ways to suppress the appetite besides the enzymes that can prevent all this overreacting in the digestive tract. First, however, you must determine your body type to be sure you have the correct information to proceed. What works for one body type may not work for another. There is one thing that should be followed by all body types who desire to burn fat and rid their bodies of excess pounds. Limit your sugar and fat intake, especially when they are combined, as in desserts. Studies have shown that you are much more likely to eat sooner after a meal heavy in sugar than after one high in complex carbohydrates or protein (plant or lean meat). My observations have led to the conclusion that most people have problems breaking down fats and handling the huge amounts of sugar we consume.

Most of my clients succeed and yet some fail. There is no magic potion that makes people lose weight and keep from gaining it again. Those who do manage to create realistic weight loss goals and meet them have several things in common.

- **They do not face any genetic obstacles.**
- **They value their mission enough to be dedicated and loyal to it.**
- **They leave out the intensity and zeal so they can enjoy themselves while making changes.**
- **They expect to succeed because they understand what works for them!**
- **The ones who do succeed not only feel good about themselves, they feel good.**

Are the foods you love an intolerance or an addiction?

For many people, the very foods they love or crave are the ones to which they are either allergic or addicted. A term you often run across in environmental medicine is "allergy/addiction." Those foods that make us unhealthy actually make us feel better for a short time after we eat them. As with all addictions, we feel better immediately following a "fix" but later we feel much worse. This means that when we give up our addictive foods we may feel ill for a short time. For instance, a carbohydrate addict more than likely will experience headaches and fatigue when the addictive foods are given up. I did not suffer the way some do when I changed my diet and gave up addictive foods. The reason for this is I had access to large doses of amylase and the supporting enzymes.

Body Type Ones will find that when large quantities of enzymes are consumed, especially a formulation including amylase, they will notice a marked decrease in their cravings for sweets. To achieve this, the formulations should contain at least 12,000 DU of amylase.

Enzymes work better when formulated together. Only the protease enzyme can stand alone. As with herbs, combining them makes them stronger, more efficient and gives them a broader range of uses. Enzyme formulations that contain only one enzyme are not desirable. The exception is protease, but only if it is taken as therapy, rather than for digestive purposes. All high dosages require taking multiple capsules on a regular basis.

Body Type Two will find the desire for salty versus sweet tastes drastically reduced with the support of lipase. The ideal formulation is 3000 LU of lipase with amylase, protease, and cellulase. You may need to take several capsules per day to equal the 3000 LU of lipase required to eliminate the sugar/salt/sugar cycle. The FDA is working on a viable way to measure lipase. For now, it is in LU, but some of the older formulations are still calibrated in UA (units of activity). In enzyme therapy we use LU as a measurement and you can get the therapeutic dosage of 3000 LU from a health-care professional.

Body Type Threes can use protease to great advantage. It works better for them in eliminating water retention than for the other types. Used in high dosages, it will cut the desire for either fatty or strong smoked meats. The mealtime formulation should include at least 50,000 units of protease activity to assist digestion. Again, several capsules will have to be taken at one time.

Protease is the "big gun" of enzymes. It offers the most benefits of all the enzymes known to us. If it is used as therapy and taken between meals, a dosage of 375,000 HUT should be taken at one time.

Body Type Four requires a multi-enzyme formulation. People in this type will often overindulge in dairy products because they find them the only foods that agree with them. With enzymes, they have more food choices and they can more easily tolerate dairy products. Type Four requires the enzymes between and with meals for optimum response. High levels of lactase and lipase need to be included in this formulation.

I have found that all four types do very well when given enzymes with the food at the beginning of the meal to assist with digestion. Taken between meals, they feed and fortify the system.

Change Old Habits

Body Type Ones will feel better if breakfast includes a protein. If they eat fruit or toast for their first meals, they are beginning their day with sugar! Do not make this mistake. Breakfast is the meal that starts this body's engine for the day. Eating protein will give you constant energy rather than a short burst followed by a long letdown. For vegetarians, eat the highest plant protein available, such as tofu. An egg is great for this body type.

All body types should learn to drink water. This should be between meals, as well as with them.

Body Type Two usually does not desire breakfast, except for a cup of coffee. Following the coffee, a slight headache and nausea can occur. Yet this body type does this day after day. If you skip breakfast, your metabolic rate is 5 percent slower than if you had eaten something. Over the course of a year you can put on eight extra pounds for not eating! Dr. Dick Couey, a professor at Baylor University, teaches that skipping breakfast can cut off five years of our life.

Body Type Three insists that breakfast is important. However, those of you in this category should not choose sausage and eggs to start your day. You already know that you will suffer gas all day long if you have a protein breakfast. A light meal of fruit and grain is best for you to start your day.

Body Type Four does best on protein in the morning, to feed and fortify the endocrine system. Breakfast is one of your major meals of the day. If you have only dairy products, however, you will be tired before lunch, and the fatigue may last all day. If you want to become a vegetarian please become a wise one. Consider fish and soybean products.

All body types should learn to drink water. This should be drunk between meals, as well as with them. However, the largest amount of water should be consumed between meals to help carry away the body's waste. I mean water, and I do not mean juices or diet drinks. A minimum of six to eight glasses per day is required. For those trying to burn fat and lose weight, a minimum of eight glasses of water in a day needs to be taken. You may be unaware that for every pound of fat burned, 22 ounces of water is released. It is extremely important to flush this water away, because it is full of the by-products of fat breakdown. This fluid is retained in the tissues, ankles, hands, and feet.

Female and Male Weight Loss Patterns

When a wife and husband come to my office for nutritional information on weight loss, I find myself giving two separate sets of guidelines. I educate them on the differences they will encounter while losing pounds and inches. Many times they are two separate body types, in addition to being two different sexes.

These differences play a big role in how they will experience weight loss and achieve their goals. Most of us have thirty billion fat cells and can store one hundred fifty pounds of fat. This applies to both male and female. Women tend to store more fat than men. Where we carry that fat depends on our body types. These factors may make it more difficult for some to successfully lose weight than for others. Please do not be discouraged by this. I did not say it is impossible. You can do it if you persevere.

Enzymes are necessary for carrying fat to our cells for storage. Conversely, they release fat from cells as well. There are two metabolic enzymes that are involved with this process. Lipogenic enzymes help the cells to store fat. Lipolytic enzymes decompose fat.

It is said that women have fatter cells because we have more lypogenic enzymes to store fat. This is because of estrogen, the hormone that makes us female. Estrogen not only stimulates the lipogenic enzymes to store more fat, but also directs it to the parts of our bodies where we accumulate fat because of our body type. According to Debra Waterhouse, M.P.H., R.D., women who take estrogen as a hormonal aid will experience an increase of fat storage.

Calories are necessary for creating fuel in our bodies. When we ingest more calories than what our body needs, the surplus is turned into fat and stored in our cells and tissue. Undigested food adds to the toxicity in our system and puts even more fat into our cells where it is stored.

Our muscle structure is determined by genes and body type. Men have 40 percent more muscle than women. Muscle contains minute, rod-like structures called mitochondria that convert fat calories to heat and water. If we have more muscle than fat cells, naturally we can burn more calories. Hence, men store less fat than women and for the most part have an easier time losing weight.

Dieting tends to double the amount of the lipogenic fat-storage enzyme. Women have fewer lipolytic enzymes, or those that release fat. Dieting can cut the number of them by half. Every time we try to lose weight we increase our metabolic enzymes that retain fat and cut back on those that help us lose it. All this has a cumulative effect. Not only do the fat cells become bigger and stronger, the muscle cells become smaller and weaker. A very important fact to always keep in mind is fat cells will not starve but muscle cells will!

Muscle is metabolic tissue that requires calories in order to survive. The more muscle you have, the greater your caloric needs. To avoid starvation, your body will grudgingly give up some of its muscle mass. However, it will do this only to provide energy for itself and to reduce the rate of metabolism so there will be a corresponding reduction in the number of calories needed for survival.

Every time you go on an unhealthy or fad diet this happens to you. The fat loss is not permanent, but the muscle loss may be. Remember, diets multiply your fat cells and make them bigger. The larger the fat cell the more likely it will survive the next diet. This enables it to divide and thereby multiply, vastly increasing your body's ability to store more fat!

Dieting does not work because we will be able to store more fat after the diet. We will have more fat cells, less muscle mass, repress our metabolism and gain more weight. Throughout all this, we have managed to sabotage our immune system and we are clinically depressed. This sounds very discouraging, doesn't it?

There must be a food plan that fits your life style and personal values. Yet, we have already learned that there is not a perfect diet. There are, however, perfect foods. These are raw fruits and vegetables that have their enzymes intact. Always utilize your body type and its guidelines and limits when choosing the correct weight loss plan for you.

Loss of inches is more important than loss of pounds. This is true because inch loss is more permanent. I suggest to my clients that rather than weigh themselves when incorporating new food choices into their lifestyle, they take measurements of their waist, hips, thighs, and upper arms. Included in these important food choices are enzymes, of course. Any real and lasting loss of fat will take place very slowly.

Women retain water and gain weight during menopause, PMS, while using birth control pills, estrogen replacement, and pregnancy. In each of these cases, hormones are involved. Men, however, produce a hormone called testosterone that influences lipolytic metabolic enzymes. These release fat, sending it to muscles instead of storing it in the cells. This is the primary reason why women lose weight differently from men. The complexity of the whole scenario is due to the specific needs, differences, and lack of enzymes of the four body types. I contend that the supplementation of enzymes with the correct foods for our individual body type will make the difference between success and failure in a weight loss program. Further and more important, the correct use of foods and plant enzymes will bring the body back into a state of balance.

At the beginning of this section I spoke of counseling husbands and wives with their expectations for weight loss. A couple whom I recently worked with comes to mind. They were both in their thirties. The woman had been taking birth control pills for five years. She had taken an assortment of diet pills as well, to take off the excess weight she was consistently gaining. Her husband was very active in sports. However, he had injured his foot and was sidelined for several months. As a result, he too had gained extra weight. The wife put him on her diet pills and by the time I saw them, they were both out of control. Depression had set in and it was affecting their marriage.

It was a challenge for me to explain to him that he was a Type Three, and what that meant for him in terms of weight loss. I advised him if he wanted to lose weight and reduce his stress levels, he ought to consider becoming a semi-vegetarian and give up red meats. The plan I suggested for his wife was very different. As she was a Body Type One, her vegetarian diet had to give way to more proteins. In both cases, I had to suggest a way for them to give up the diet pills, a harmful stimulant.

This young couple left our consulting session armed with information about enzymes and a body type booklet. Their next appointment was two weeks later. They were pleased to report that they were getting to know each other better every day. The differences they shared, and how each one reacted emotionally and physically was a great marvel to them. This all took place three years ago. I realized as I wrote this chapter that this couple is still vigilantly taking their maintenance enzymes. I received a phone call from them last December. They called to say they are doing beautifully and wanted to know if they should take extra enzymes during the Christmas holidays. Working with people who experience such positive changes in their lives is a constant source of joy for me.

There are weight-loss stimulants offered in every magazine on the market today. In truth, some of them work temporarily. When the initial effects wear off, the users simply increase the dosage. At last they realize the pills will no longer do the job, but by that time they are ready to crash and burn from anxiety and depression. Our bodies will not tolerate continuous stimulation. When the metabolism speeds up, the body reacts. After about twenty-one days the body makes another attempt to protect itself. It slows the metabolism so it can restore homeostasis. The human body fights very hard to avoid compromising its vital functions.

Body Slimming Exercises For Each Body Type

Your personalized plan depends on your major body type.

Body Type One will find that swimming and low-impact aerobics are excellent forms of exercise for them. However, only two or three times a week for twenty-minute periods are best. Walking is the most natural way for your body type and can be done daily for short periods of time. Building endurance slowly is best for this type. This will free your body of fatigue and energy swings you frequently experience. Strenuous or intense exercise is not good for your balance. Balancing your body will help you to feel happy and gain a sense of freedom.

Body Type One women will collect cellulite in the hips and thighs. Men collect it in the midriff bulge at the waist. Changes in diet, enzymes, and appropriate exercise will rid your body of cellulite and fat from the midriff down to the thighs. Your energy level will improve and your mood swings dissipate. Gradually you will achieve overall equilibrium.

The best oils to use for Body Type One are lavender, peppermint, rosemary, and juniper.

Suggested Oils: There are many good oils on the market. The following oils are good for balancing the four body types. Use of these essential oils and the various blends are the result of many years of aromatherapy work and study by professional aromatherapists. The oils suggested here are specifically designed to alleviate stress that accompanies each individual body type. At the end of each body-type discussion, the oils best suited for you for both massage and bath are listed. These oils are not intended for digestion and should not be taken internally. They are for external use only.

Taking oil baths to help create balance is highly recommended for all four body types. Blended oils can be used in the shower. However, bathing with oils for twenty minutes in water that is not too hot is best. Using these oils in an aroma diffuser or aroma dish to add fragrance to the air you breathe works very well to balance the senses.

Massage Oils: To obtain the correct blend of oils for your body type, you may use as many as three of those listed. A maximum of twenty drops per one ounce of carrier oil is allowable. For example, choose your three favorite oils from the list. These will be the ones that feel or smell the best to you. Use from four to six drops of each and blend them with the carrier oil of your choice. Almond oil is a great carrier and very nourishing. Jojoba oil comes the closest to natural skin oil.

Bath Oils: For a perfect bath, run the water and fill the tub before adding the oils. If the oils are added to the running water, they evaporate. The suggested amount to use is six to eight drops of oil per bath. Less than that is recommended if you use the more stimulating oils like cinnamon, lemon, spice, or peppermint.

Body Type Two will discover that power walking and normal walking or jogging are the best ways to exercise. These kinds of exercises can be enhanced by carrying or wearing light wrist weights. Two other forms of exercise that will serve you well are dancing and gymnastics. The primary goal for you is to bring balance to your upper and lower body simultaneously.

If you are a Type Two you can improve the circulation below your waist with a foot vibrator. This can be purchased at health food and department stores. Electric massagers are all right if you use them on the entire leg. Do some exercising two or three times per week. Finish each session by elevating the legs for a few minutes, or massaging the lower part of your body. Balancing your body type will make you understand that you can accomplish anything you want to do!

When your body type is lacking digestive balance it stores the toxins in your fat reserves. Cellulite is toxic fat. You carry cellulite in your buttocks and upper thighs. Enhancing your life with changes in diet, enzymes, and appropriate exercise will dissolve the fat and cellulite.

The best oils to use for Body Type Two are jasmine, geranium, sage, and juniper.

Body Type Three will dance, play tennis, or racquetball for ideal exercising. One of these should be done at least two or three times a week. If none of them appeal to you, try walking. However, if you choose walking over the other three, you must walk every day for thirty minutes or more. You require a full body exercise. Weight lifting, though, is not the best choice for you. It may even contribute to your state of imbalance. You need to become more flexible and rid your body of toxic congestion.

Rather than storing your toxins in cellulite the way Types One and Two do, yours accumulate in the muscles and colon. This creates body aches, shoulder pains, and gas or a bloated sensation. If you should begin to collect cellulite, it will be later in life. Your cellulite will accumulate on your back and upper arms as well as your stomach. Massage is very good for your body type; one per week would be ideal. Once you have brought your body back into balance, you will feel less stressed and more in control.

The best oils to use for Body Type Three are ginger, eucalyptus, geranium, and juniper.

Body Type Four will find aerobic dancing and martial arts, especially Tai Chi the best forms of exercise. Walking is all right, but it is not the best choice because you think and obsess too much. An exercise like walking allows you to continue doing this. You need something that is not mechanical, something that requires focus and concentration so your mind cannot run wild. Yoga is a good possibility if you are up to the challenge. The Type Four must empty the mind, and can do so with the practice of meditation. Do one of these kinds of exercise two to three times a week. The goal for you is to exercise from the neck down. Try getting a massage at least twice per month. This would be very good for you. As you approach a more balanced state, you will feel peaceful and much more in control of your life.

When the Type Four is lacking in digestive balance, he or she will store cellulite under the skin throughout the body, particularly in the knees and breasts. This is what gives you a soft or baby fat look. Your changes in diet will include dropping dairy products. A good plant enzyme regimen and appropriate exercise will assist you in your quest for balance.

The best oils to use for Body Type Four are patchouli, ylang-ylang, sandalwood, and juniper.

A Personalized Plan for You: Body Type One

A typical Type One is born with specific characteristics such as body shape, food cravings, and personality traits. If you are a Type One, you are aware of the powerful energy and drive that you feel when you are stable and secure. You are very intuitive and highly thought of for your wit and caring attitude toward others. When you are in a healthy, balanced state you are very much in control of your world. When you are out of balance, you typically experience cold hands and feet, skin problems, hypoglycemia, and nervousness. You may have an inability to utilize vitamin C or properly deliver calcium. Your primary enzyme deficiency is amylase and, over time, you may develop a shortage of protease as well. The following page shows a list of energy-producing foods for your body type.

You may be surprised to see that some of your favorites are not listed. The missing items represent foods that do not help you to create balance in your body type. For instance, carrots are a food that is too high in sugar. Unless they are cooked, they can create gastrointestinal distress for you. Vegetables not included should be steamed if you must eat them. It is preferable, however, to simply avoid the items that are not listed. All fruits should be ripe. Bananas are not included because of the high sugar content and gastrointestinal distress they cause for your body type.

ENERGY-PRODUCING FOODS FOR BODY TYPE ONE

VEGETABLES (should be eaten in a raw state whenever possible):

Asparagus	Celery	Okra
Broccoli	Cucumber	Peas
Brussels sprouts	Green beans	Sprouts
Cabbage	Leafy greens	Sweet peppers
Cauliflower	Mushrooms	Zucchini

LEGUMES:

Soybean products	Tofu
Mung beans	Chick peas (garbanzos)

FRUITS:

Apples	Grapes	Pears
Avocados	Lemons	Pineapple
Cherries	Mangoes	Plums
Figs	Melons	Prunes
Grapefruit	Oranges	

GRAINS:

Barley	Oats
Wheat	Brown rice

DAIRY:

Cheese (white and soft)		Yogurt (plain)

NUTS:

Sunflower seeds	Almonds	Most kinds of nuts

Nuts should always be roasted to remove the enzyme inhibitors.
Peanuts are the most difficult to digest.

OILS:

Olive

PROTEINS:

Fish	Poultry	Eggs
Beef	Soybean	Wheat
Brown rice	Black beans	

FOODS TO AVOID:

Coffee	Tea	Sugars
Desserts	Fruit juices	Fried foods
Heavy oils	Refined grains (white flour)	

NOTE: If you have had problems in the past with some of the foods listed here, plant enzymes will greatly assist your digestion.

Sample Daily Diet: Body Type One

Breakfast: Eggs or choice of plant proteins, one piece of raw fruit, raspberry tea. If you want toast eat a high complex-carbohydrate bread such as seven-grain or protein bread. It is very important to start your day with protein.

Lunch: Large green salad with dressing, choice of poultry or fish, brown rice, and raspberry tea. One piece of raw fruit for dessert. Eat only raw fruit, and only one selection.

Dinner: Choice of chicken, turkey, or fish. Raw salad and steamed vegetables. Limit your bread and potato intake.

Enzyme Therapy: Body Type One

Take a multiple digestive formulation containing at least 10,000 units of amylase plus ample lipase, protease, lactase, maltase, and sucrase in each capsule. You will take the capsules in a quantity that is relative to the amount of food you are eating. For example, the meals I have suggested require two to three capsules per meal, depending on the strength of the formulation.

The antioxidant formulation includes at least 375,000 units of protease to cleanse and fortify the liver and blood. Take these between meals, not with food. You will need to take several capsules per day.

A glandular and enzyme formulation would be good for you to fortify the endocrine system and to work on your hypoglycemia. Take these after eating or at any other time that suits you.

A bacterium product to balance the pH of the large and small intestines would be ideal, since your body type has a weakness for fungal forms.

Body Type Two

You were born with specific characteristics. The way you are shaped, your food cravings, some personality traits—all these added together equal you as a unique human being. When you are feeling well, you are very self-assured, full of enthusiasm and creativity. You are naturally sensitive, and when you are in a state of balance you have great confidence in your world and yourself. Type Two learns very quickly. You use your wonderful imagination to make the world a better place for everyone. When you are out of balance the health problems you are most likely to experience are allergies, indigestion, arthritis, skin conditions, and weak kidneys. You are deficient in lipase and may be unable to utilize fatty acids, amino acids, hormones, and vitamin B. The list on the following page suggests energy-producing foods for your body type.

Sample Daily Diet: Body Type Two

Breakfast: Fruit from your list, a vegetable protein such as scrambled tofu, and herbal tea.

Lunch: Green salad, vegetables, cottage cheese, eggs, or plant protein. One slice of whole grain bread and a piece of fruit.

Dinner: Choice of chicken, turkey, or fish, brown rice and raw fruit.

Fats and spices overstimulate your gallbladder. If you have skin problems, they can be controlled with this diet.

Enzyme Therapy: Body Type Two

Use a multiple digestive formulation containing at least 3000 LU of lipase, with ample amounts of amylase, protease, cellulase, sucrase, maltase, and lactase in each capsule. Take them according to the size of the meal you will eat. If you are bloated or have indigestion, take the capsules immediately before and directly after the meal. In addition you should take lipase between meals for high cholesterol or triglycerides. A high dosage of 375,000 HUT protease between meals to build the immune system. Before bed a bacterium product preferably containing *L. Plantarium* for good colon health.

ENERGY-PRODUCING FOODS FOR BODY TYPE TWO

VEGETABLES (should be eaten in a raw state whenever possible):

Asparagus	Garlic	Onions
Beets	Green beans	Radishes
Carrots	Okra	Turnips
Cucumber		

Steam the following:

Broccoli	Eggplant	Potatoes
Cabbage	Mushrooms	Tomatoes
Cauliflower	Peas	Zucchini
Celery		

LEGUMES:

Mung & Kidney beans	Chick peas	Red lentils

FRUIT: (1)

Apricots	Grapefruit	Oranges
Bananas	Grapes	Papaya
Berries	Lemons	Pineapple
Cherries	Mangos	Plums
Dates	Nectarines	Tangerines
Figs		

GRAINS:

Oats	Rice	Wheat

DAIRY: (2)

Yogurt	Cottage cheese	White & soft cheeses

NUTS: (3)

Pumpkin seeds	Sunflower seeds	Walnuts

OILS:

Olive	Safflower	Sesame

PROTEIN:

Poultry	Fish	Beef
Eggs	Tofu	Brown rice & beans

FOODS TO AVOID:

Spices (4)	Creams	Organ meats
Butter	Rich desserts	Pork
Fats	Ice cream	
Fried foods	Most oils	

NOTES: (1) Eat apples sparingly, they may cause gastrointestinal discomfort. (2) Always take digestive enzymes when eating any dairy products. These items produce mucus and allergies without the proper enzyme action. (3) Nuts are acceptable in small servings. Be certain they are good quality and take a digestive enzyme with them. (4) Always use spices in moderation. Minimize or avoid dill, saffron, turmeric, parsley, and coriander seed.

Your antioxidant formulation must contain a minimum of 375,000 units of protease for cleansing and fortifying the system. These are to be taken between meals. For the colon, take a combination of enzymes and herbs.

Add to this a good bacterium product. In all body types, it is very important to see a health-care professional for advice so you can be sure you are taking the correct formulations.

Body Type Three

You are born with traits that are unique to your body type, just as the other three types are. Your inherent body shape, food cravings, and personality are examples of these characteristics. When you feel you are in control of yourself and your world, you have immense power and drive. You are very perceptive. When you are in a state of balance, you can be a leader in whatever you choose. The common disorders you suffer when you are out of balance are exhaustion, body aches, high blood pressure, constipation, and insomnia. Your signature enzyme is protease. If it is lacking, you may be troubled by kidney disorders or a build up of uric acid that may result in gout and water retention. Sometimes, you may be unable to utilize vitamins C and E and calcium. The following is a list of energy-producing foods for your body type.

ENERGY-PRODUCING FOODS FOR BODY TYPE THREE

VEGETABLES (should be eaten raw whenever possible):

Asparagus	Celery	Onions
Beets	Eggplant	Peas
Brussels sprouts	Garlic	Peppers
Cabbage	Leafy greens	Potatoes
Carrots	Mushrooms	Radishes
Cauliflower		

LEGUMES: All are acceptable except tofu. Eat kidney beans in moderation.

FRUITS:

Apples	Cranberries	Pears
Apricots	Pomegranates	

Be careful of very juicy fruits like grapefruit or other citrus fruits.

GRAINS:

Barley	Corn	Rye
Buckwheat	Millet	

Hot cereals and steamed grains are too moist and heavy for you. They will create a bloated feeling unless taken with enzymes.

DAIRY:

Milk	Yogurt	Cottage cheese

Always choose the lower fat dairy products. They are difficult for you to digest without enzymes.

NUTS:

Sunflower Seeds	Pumpkin Seeds	Walnuts

Reduce your intake or completely avoid all others. Protein is difficult for you to digest.

SPICES: Use salt sparingly, particularly if you are accustomed to using it at every meal.

OIL:

Safflower	Canola	Olive
Sesame		

PROTEIN: (1)

Poultry	Fish	Eggs (3/wk max.)
Tofu	Brown rice & beans	

FOODS TO AVOID:

Salty foods	Salt	Red meat
Pork	Organ meats	White flour
Fried foods	Heavy oils	

NOTES: (1) Eat eggs only occasionally. This is not because of the cholesterol, but because of your difficulty with digesting protein. A semi-vegetarian diet is ideal for most of you. If some of these foods have been troublesome for you in the past, plant enzymes will help you with digestion. That should be sufficient to take care of any problems you may have had.

Sample Daily Diet: Body Type Three

Breakfast: Light cereal or seven-grain toast, raw fruits, decaf coffee, or herb tea.

Lunch: Large green salad, fish, and fruit. Do not use a creamystyle salad dressing. You may prefer a vegetable protein rather than fish.

Dinner: Choice of chicken, turkey, or fish; steamed or raw vegetables and raw fruit. This is the meal at which you will have the easiest time digesting your protein.

Enzyme Therapy: Body Type Three

Take a multiple digestive formulation containing at least 50,000 units of protease activity, plus appropriate amounts of amylase, lipase, lactase, sucrase, maltase, and cellulase in each capsule. The dosage should reflect the quantity of food you are eating. This depends too, on the strength of the formulation, especially the protease.

You need an antioxidant enzyme formulation which must contain at least 420,000 units of protease activity to fortify the liver and blood. Protease is the enzyme that is most important for your body type. Take this regimen several times per day between meals.

A bacterium product to keep the pH in the small and large intestines constant should be taken before bed. This helps prevent gastrointestinal problems.

Body Type Four

You are born with specific traits like your body shape, your food cravings, and the assorted qualities of your personality. When you are enjoying a state of good health and stability, your mind is very clear and creative. You tend to be a reflective, logical thinker and visionary. As long as you maintain balance, you are in complete control of your world and you overflow with helpful information for others. Your out-of-balance symptoms are allergies, milk intolerance, aching knees, gas, fatigue, nervousness, skin breakouts, and irritable bowels. The enzyme deficiencies most commonly occurring in your body type are lipase, amylase, and lactase. This means your needs are more difficult and complex than those of the other body types. You possess an inability to assimilate fatty acids, amino acids, and fiber. The list on the following page includes the energy-producing foods best suited for your body type.

Sample Daily Diet: Body Type Four

Breakfast: Have a substantial meal with protein. It is extremely important for you to begin your day with protein. Have this with some raw fruit. Be careful with cereal and milk; they throw you out of balance.

Lunch: Choice of light poultry, fish or eggs, steamed vegetables, and one piece of ripe fruit.

Dinner: Choose another protein, more steamed vegetables, a whole-grain roll, herbal tea, and raw fruit for dessert.

Enzyme Therapy: Body Type Four

Take a multiple digestive formula containing at least 10,000 units of amylase, 500 units of lipase, 30,000 protease, 1000, and lactase. If you eat dairy you will need additional lactase at 8000 ALU.

You need an antioxidant enzyme formulation which must contain at least 420,000 units of protease activity to fortify the liver and blood. Take this regimen several times per day between meals. A bacterium product taken at bedtime should help prevent irritable bowel syndrome.

ENERGY-PRODUCING FOODS FOR BODY TYPE FOUR

VEGETABLES: Although the other three types do better with raw vegetables, you should steam yours. This is why some of you suffer with indigestion and gas after eating salads. I am aware that steaming may kill the enzyme action in these vegetables, and therefore it would seem to be better to eat them in a raw state. Tolerating raw foods can be difficult because your digestive system is very sensitive.

Asparagus	Green beans	Onions
Cauliflower	Leafy greens	Peas
Celery	Mushrooms	Spinach
Eggplant	Okra	Tomatoes
Garlic		

LEGUMES:

Mung beans	Tofu	Soybean products
Garbanzos	Kidney beans	

FRUIT:

Apples	Figs	Plums
Grapes	Lemons	Tangerines
Grapefruit		

Fruit must be very ripe and eaten raw. Avoid canned or frozen fruits.

GRAINS:

Buckwheat	Barley	Toasted grains
Brown rice	Oats	

DAIRY: Dairy products act as a stimulant to Type Four and should be taken sparingly. They are the most difficult for you to digest and they create mucous and allergies. If you eat them, be sure to take enzymes.

PROTEIN:

Poultry	Beef	Fish
Eggs	Tofu	

NUTS:

Almonds	Coconuts	Cashews

Most nuts if roasted.

OILS:

Virgin olive oil	Sesame

FOODS TO AVOID:

Ice cream	Sour cream	Cheese
Butter	Sugar	White flour

Fried foods and all other dairy products.

NOTES: Protein from dairy sources is your big problem area. Therefore, a good alternative for you would be fish, poultry, and eggs. It is important that you obtain amino acids in food form. Spices should be used cautiously. Please remember, these are not the only foods you can eat, but they are the best ones for keeping your body type in balance. Plant enzymes are always very helpful for your digestive process.

At the beginning of this book I explained that the body uses enzymes to operate every action, function, and process it needs to stay alive and healthy. Enzymes deliver the forty-five necessary nutrients for fueling the body and creating new cells. The ability of your body to digest, assimilate, utilize, and eliminate food is dependent on digestive enzymes. The availability of these enzymes to you is known as your enzyme potential. The most important factors involved in enzyme potential are heredity and the strengths and weaknesses of your individual organs, glands, and cells.

Stress, too many cooked foods, aging, your environment, and various other elements too often join forces to create a lack of these life-giving enzymes. In this book, I have given you a detailed and accurate picture of what can happen when we experience enzyme deficiency. Together we have explored what it looks and feels like to be in a state of less-than-perfect health. Life cannot exist without enzymes. We all know that without good health, life can be a burden. If you have understood this and gained something from that knowledge, then my work is done. Writing this book has been worth every moment of time and every bit of effort I have put into it.

Part IV

Enzymes and Your Pet's Health

Chapter 11

The Importance of Enzymes in Your Pet's Diet

Our pets don't deserve to eat one meal without enzymes. When you consider the foods that we feed our pets are processed, canned, cooked, and enzyme damaged in some way. Remember, heating food above 118 degrees will destroy the enzyme activity so it is no wonder that our pets have all the same digestive disorders and diseased conditions that people have. In fact pets normally do not have a choice in the foods they are fed.

Dogs and cats digest their foods just as we humans do. They require digestive enzymes and their food is broken down in the same digestive pattern. We humans can tell someone when we hurt or have poor digestion, but our pets can't. It shows up as constipation, diarrhea, itchy skin, and shedding of their fur until it turns into a disease. Just like humans they inherit an enzyme potential that is dependent on their own unique DNA. Add the genetic problems in breeding of the animals and you compound their predicament.

In fact, pets normally do not have a choice in the foods they are fed.

Enzyme deficiency may be the primary reason for many of the health problems existing in veterinary concerns today. We have recently become involved with the enzyme therapy for Killer Whales and Dolphins. These mammals are fed only fish. The fish is slightly frozen and given for food. In fact, they are treated royally and given the best of care. Yet, they become ill and have low immune systems and all other medical possibilities that can take place amongst their breed. It has been exciting and very fulfilling to have a part in their health and recovery and we are using pharmaceutical grade plant enzymes. The response has been wonderful and why not? Plant enzymes work throughout their entire digestive tracts as well as in humans.

Nature intended for the enzymes in food to assist the natural digestive enzymes secreted by the animal. If food is not digested as nature planned, it cannot supply an animal with the nutrition needed to build and repair vital organs and tissue, produce energy, and maintain a strong immune defense system.

Enzyme synergy and cell metabolism are important for complete utilization of the potential nutritive value of foods. In the absence of active enzymes in food, the foodstuffs are still able to be digested and the nutrients are still released from the food, *but not at maximum efficiency*. Evidence of this statement is the observation that food undergoes a decrease in nutritive value, in addition to the well-known loss of vitamins, when cooked and /or processed.

Digestion is the basic process in which specific enzymes break down protein, carbohydrates, fat, and fiber so that these food nutrients can be utilized by the body to maintain good health.

Research has shown animals in the wild have much less chronic degenerative disease than domesticated animals. This can be directly attributable to their enzyme-rich food sources. Cooking or processing of food destroys its enzymes. When animals are fed cooked, canned, dried, or any of the processed food stuffs, the responsibility for digestion falls entirely on the animal's digestive system. This compromises their immune system. When a multi-enzyme formulation is added to the animal's diet, the same enzymatic activity is available for digestion that could have been provided by raw food enzymes. This relieves some of the burden placed on the digestive and immune systems.

Research has shown animals in the wild have much less chronic degenerative disease than domesticated animals.

When the digestive system becomes overburdened, the body begins "stealing" enzymes from the immune system, leaving the animal at a higher risk for disease or illness. Therefore, enzyme supplementation to the pet's diet will ease the burden on its digestive system.

An enzyme-deficient diet may lead to musculoskeletal inflammation, a major cause of discomfort and debility in millions of older canines. Osteoarthritis, loss of mobility and movement, pain, and swelling can be relieved with enzyme supplementation. Relief from edema, a major source of irritation for an animal, has also been experienced.

Protease enzymes break down protein food stuff. When taken between meals protease enzymes hydrolyze large protein molecules into smaller polypeptides and amino acids. When larger protein molecules (bacteria, cellular debris, parasites, and fungal forms) are hydrolyzed and become soluble, they are more readily passed in the urine.

Lipase enzymes aid in proper fat absorption. When taken between meals it also provides essential fatty acids to maintain skin, coat, and other tissues.

Amylase enzymes assist in the breaking down of carbohydrates. If carbohydrates are not properly digested they cannot be used for energy. When taken in between meals they aid in the control of inflammation.

Cellulase enzymes assist in the breaking down of fiber and other nutrients. Taken between meals, cellulase controls malnutrition and bowel disorders.

Case History

Dr. Jim Smith a veterinarian specializing in animal allergy and dermatological problems has been performing clinical research with plant based enzymes. He is thrilled with the results and says the time expected for healing response is not just cut in half but in many cases in 72 hours. This of course depends on the condition of the animal. But in all the years of his practice he has not seen anything in medicine or any other supplement work like the enzyme protease for mange, foot biting, mites, parasites, ear infections, and various other conditions that he sees on an ongoing basis in his clinics. Dr. Smith has been working with other veterinarians by teaching veterinarian enzyme therapy usage.

Vets using the enzymes have seen amazing results with large animals such as horses and even the very large elephants. The challenge is for me to come up with some of the dosages so it has been a real education. Most all of the doctors tell me that 50% and higher of the ailments that bring animals to their clinics are directly related to diet and nutritional imbalances.

What enzymes have done for the health of the animal kingdom is exactly the same as for the human race. I know that by now you surely have some usable information about enzymes for the health of your family and your pets. But please remember the best way is the preventative way. However, if you have a condition it is important for you to educate yourself and look at all possible choices.

Part V

Questions and Answers

Chapter 12

Questions and Answers about Enzymes

Dr. Dick Couey is a Professor of Nutrition and Health Sciences at Baylor University School of Medicine Waco, Texas. He is an eminent researcher and author of thirteen books on nutrition and other health related subjects. I had the good fortune to spend time with Dr. Couey recently. This is what he had to say about enzymes.

> **All my life I have been searching for answers that would allow us to obtain good health, free from hypokinetic [a condition of abnormally diminished muscular movement] or degenerative diseases. First, I have studied how exercise changes the physiology of the human cell. I found that proper exercise will definitely improve the physiology in the cell. However, I found that exercise alone will not protect the cell from dysfunction. Secondly, I studied how nutrition affects the cell. I found that eating the forty-five known nutrients in their proper amounts will definitely improve the inner workings of the cell. However, I found that proper nutrition alone will not completely keep the cell from abnormalities. Up to this point in my life, I felt that exercise and proper nutrition were the best remedies to protect the cell from irregularities. I did not have the answers when some of my fit friends experienced health difficulties. Now, through the introduction to Dr. DicQie Fuller and her knowledge of enzyme nutrition, I have discovered the answer to good health. Enzymes are the answer. Enzymes are needed for every chemical action and reaction in the body. Our organs, tissues, and cells are all run by metabolic enzymes. Enzymes are the labor force of the body; without them the forty-five known nutrients cannot be fully utilized by the body. Our cells are an orderly, integrated succession of enzyme reactions. Without enzymes, it does not matter how much exercise you**

get or how well you eat, your health will be severely affected. Raw food naturally contains enzymes to help digest that food. Cooking of our foods destroys the enzymes that we would have received from the food, forcing our bodies to produce all of the enzymes needed to complete the digestive process. This depletes the enzyme potential we inherited. When the body's ability to produce enzymes is weak, our health will be affected. Hopefully, from reading this book you will gain some knowledge about how enzymes can insure your good health, and protect or rehabilitate you from poor health.

Dick Couey, Ph.D.
Baylor University, March 1996

The following questions and answers have been compiled by Dr. Dick Couey and me, and represent the most commonly asked questions we both hear about enzymes. They are excerpted from the book we have co-written on nutrition and enzymes titled *Living Longer: Questions you never know to ask, answers you can't live without.*

DicQie Fuller, Ph.D.
Houston, Texas, March 1996

What are enzymes?

Enzymes are protein molecules that carry a vital energy factor needed for every chemical action and reaction that occurs in our bodies. There are approximately 2,700 different enzymes found in the human body. These enzymes can combine with co-enzymes to form nearly one hundred thousand various chemicals that help us to see, hear, feel, move, digest food, and think. Every organ, every tissue, and all the one hundred trillion cells in our body depend upon the reaction of enzymes and their energy factor. Nutrition cannot be explained without describing the vital role played by enzymes.

Can you simply define nutrition?

Simply stated: Nutrition is the body's ability to consume the forty-five known nutrients in their proper amounts, digest these nutrients, absorb them, carry them into the cells, metabolize these nutrients, and eliminate the waste without getting fat. The following is a list of the forty-five known nutrients.

- **Carbohydrate**
- **Lipids (fats)**
- **Protein**
- **Water**
- **9 Amino Acids**
- **13 Vitamins**
- **19 Minerals**

Eating these foods, (which includes ingesting their enzymes) in their proper amounts, will normally ensure good nutrition. Enzymes are responsible for digestion, absorption, transporting, metabolizing, and eliminating the waste from these nutrients. Again, every organ, every tissue, and all of the one hundred trillion cells in our body depend upon the reactions of enzymes and their energy factor.

What do you mean by "energy factor"?

The energy factor is the energy that triggers or starts the chemical reactions between enzymes and other substances in the body. This energy factor is separate and distinct from the chemical makeup of the enzyme itself. A good example of this energy factor can be observed by placing a raw bean into a pot of boiling water. The cooked bean will fail to sprout. Its life force (energy factor) has been removed. Science tells us that only living organisms can make enzymes with this energy factor. Chemicals that serve as catalysts work by chemical action only, while enzymes function by both biological and chemical action. Catalysts do not contain the energy factor which is measured as a kind of radiant energy emitted by enzymes. The energy factor of enzymes has never been synthesized.

Simply stated: The energy factor is the electricity that makes the light bulb (the enzyme) work.

Are there different types of enzymes?

Enzymes can be classified into three primary groups. They are **digestive**, **metabolic**, and **food enzymes. Digestive enzymes** are secreted by the salivary glands, the stomach, the pancreas, and the small intestines. The digestive enzymes break our foods down in the gastrointestinal (GI) tract so that the forty-five known nutrients can be absorbed into our bloodstream. **Metabolic enzymes** are those enzymes that act as catalysts (speeding up chemical reactions) within the cells. These metabolic reactions can be likened to the production of energy. **Food enzymes** are those enzymes that are present in raw foods. Cooked foods lose their energy factor and the enzymes in them are rendered ineffective.

What are the names of the enzymes?

There are many enzymes. However, the most notable ones and their primary functions are:

Digest Proteins:	**Protease / aminopeptidase / enterokinase / carboxypeptidase / trypsin / pepsin**
Digest Lipids:	**Lipase / cholesterol esterase**
Digest Carbohydrates:	**Amylase / sucrase / maltase / lactase / ptyalin / cellulase**

Protease, lipase, amylase and cellulase are the four main categories of food enzymes. **Protease** digests *protein;* **Amylase** digests *starch;* **Lipase** digests *fat;* and **Cellulase** digests *cellulose* (fiber). God's natural plan calls for raw food enzymes to help with digestion instead of forcing the body's digestive enzymes to carry the entire load. Raw foods contain these four enzymes. They break down the proteins, fats, carbohydrates, and fibers contained in our foods.

How do enzymes work in our bodies?

When we eat raw foods the enzymes in the food are activated by heat and moisture in the mouth. Once active, these enzymes digest a significant portion of our food and make it small enough to pass through the villi (small projections found in the small intestine) and from there into the blood. Metabolic enzymes living in the blood take the digested forty-five known nutrients and build them into muscles, nerves, bones, blood, lungs, and various glands. Every cell in the body depends on a specific group of enzymes. Each enzyme has a particular function which is referred to as enzyme specificity. A protein digestive enzyme will not digest fat; a fat enzyme will not digest starch. Enzymes act upon chemicals and change them into other chemicals, yet remain unchanged themselves. Simply stated; our chemicals are changed from their original identity by the associated enzyme into another chemical with a different identity. Without enzymes, nothing in our bodies would work.

I've heard that enzymes are destroyed by the acidic environment in our stomach. Is that true?

Destruction of our enzymes in the stomach was a theory held by many researchers. However, recent research[4] reveals that enzymes are only *deactivated* in the lower acidic part of the stomach (pylorus). When the enzymes reach the alkaline environment of the small intestine, nearly all of them become reactivated, aiding in completion of the digestive process. When we eat cooked food, all the enzymes have been destroyed. The entire burden of digestion is placed upon the body's ability to make enzymes from the limited reserves that are stored in the body.

Does cooking our food destroy enzymes?

Unfortunately, yes. Some enzymes are completely denatured (broken up) when exposed to temperatures over 118 degrees Fahrenheit for any length of time. Enzymes cannot tolerate heat very long at all. They are different from vitamins and minerals in this respect. If water is hot enough to feel uncomfortably to the hand, it will injure enzymes in food. Most food sold in stores has been heat processed to some extent.

If enzymes are destroyed by cooking, then how do we get enough enzymes?

There are two ways to acquire enzymes—eating raw foods and by taking plant enzyme supplements. The body can make enzymes; however, the more enzymes it is required to produce for digestion, the fewer enzymes it can create to maintain optimal health. It is very important that we preserve the body's ability to make enzymes. Otherwise, the body will become dangerously depleted of this life force, resulting in serious health consequences.

How can you preserve your body's enzyme level?

It seems that we inherit a certain enzyme potential at birth. However, we cannot depend solely on our bodies to produce all the enzymes we need. In this case, our enzyme potential will be undermined much faster than intended by nature. To fortify your enzyme potential, you must eat raw foods as much as possible, or take plant enzyme supplements. Failure to do so may result in serious illness or even early death.

What happens to our bodies when we exhaust our enzyme reserves?

Research[16] done on rats and chickens that were fed cooked foods revealed that the pancreas gland enlarged to handle the extra burden of the enzyme-deficient diet. Hence, the animals got sick and failed to grow. The pancreas is responsible for making and secreting many digestive enzymes. Our pancreas will enlarge when called upon to process more enzymes or digest cooked food. Ruminant animals such as cattle, goats, deer, and sheep get along with pancreas about a third as large as ours because of their raw food diet. However, when these animals were fed heat-processed, enzyme-free food, their pancreas hypertrophied (enlarged) up to three times the normal size than when fed on a raw plant diet.

Our health suffers an extremely adverse impact when the life-giving enzymes have been cooked out of our foods. This has been dramatically demonstrated by studies[15] conducted with dogs at Washington University. Researchers drained all the pancreatic juice enzymes from the dogs. Despite being given all the food and water they wanted, all the animals died within a week. As you can see, exhausting our pancreas production of enzymes can lead to death.

What other evidence is there that proves we waste our enzymes?

Only humans can live a long time on enzyme-free food. All wild animals acquire their enzymes from raw food. Wild animals that eat raw food do not require the rich concentration of enzyme activity in their digestive juices that humans do. For example, wild animals (deer, elephants, and other ruminants) have no enzymes at all in their saliva. When we examine human saliva we find high concentrations of ptyalin (an amylase enzyme that digests starch). When dogs and cats eat their natural, raw, carnivorous diet, there are no enzymes in the saliva. However, when dogs are fed on a high carbohydrate, heat-treated diet, enzymes show up in the saliva within about one week. This indicates that we waste our enzymes on digesting cooked food. Our body has to adapt by producing additional digestive enzymes.

Other evidence that suggests we may be squandering our reserves of enzymes is that most wild animals have a smaller pancreas. This indicates that the wild animals get along with far less pancreatic enzymes than we do.

Do wild animals secrete enzymes in their stomachs?

Most of them do not. For example, cetaceans, (whales, dolphins, and porpoises) have three separate stomachs, none of which secrete enzymes or stomach acid. This brings up the question, how do these animals break down their diet of seals and fish? One answer may be that the seals and fish provide enzymes and pancreatic juices in their own GI tract. When the whale swallows the seal, the digestive enzymes of the seal become the property of the whale. These have become the food enzymes for the whale and work on behalf of it throughout the many days required to digest the food and empty the contents of the whales's stomach. In addition, all animals have proteolytic (protein digesting) enzymes known as cathepsins, which is found in their muscles and organs. After death, the animal's tissues become acidic, thereby activating cathepsin. This in turn causes the breakdown of the cells and tissues of the animal. Perhaps this will help explain why wild animals do not have enzymes or acid in their stomachs.

Can you trace how enzymes work in digestion?

Yes. When we eat raw food, enzyme activity begins at the very moment the cell walls are ruptured through mastication (chewing). After swallowing, digestion continues in the food enzyme section of the upper stomach (fundus) for one-half to one hour, or until the rising tide of acidity reaches a point where it is inhibited. Then the stomach enzyme (pepsin) takes over. High levels of acidity will deactivate the enzymes.

It should be pointed out that during the half hour or so that the food is in the upper part of the stomach, enzymes from raw food break down the carbohydrates, fats, and proteins. However, if you eat cooked, enzyme-free food, the food will stay there for up to an hour. During this time, only very minimal digestion takes place. The salivary enzymes work on the carbohydrates but the protein and fat must wait. This is an appropriate time to take the proper enzyme digestive supplements. If supplemental plant enzymes are taken with a meal, these digestive enzymes begin their work immediately. Plant enzymes will work on the protein, carbohydrates, and fat during the hour or so that the food remains in the upper part of the stomach. The supplemental plant enzymes will break down the food, thus saving the stomach from having to release as many of its own enzymes. This desirable reaction results in conservation of the enzyme potential and body energy. It allows the body to devote its attention to supplying more metabolic enzymes so the organs and tissues can carry on their daily work. All this action summed up equals preservation and maintenance of good health.

I told my doctor that I was going to take enzymes and he said they were of no use to the human body. Why would he say this?

In medical schools and dietitian textbooks, we were taught for years that enzymes are active in our bodies throughout our lives. Therefore, many people in the medical profession believe that there is no need for supplementation. Added to that, a widely held belief prevailed that enzymes would not be useful because they are destroyed in the stomach during digestion. As in any profession, we assume the information we received in the schools we attended has been diligently scrutinized and the latest research regularly examined. This is normally how we are kept up to date.

However, research and scientific papers have been submitted on the subject of digestive plant enzymes and enzyme potential for the last sixty-five years. The latest information, which is really not new at all, comes from studies done on humans rather than laboratory animals. One such study comes from the Department of Biochemistry, School of Medicine, Wright University, Dayton, Ohio, 1992–93. Doctors Prochaska and Piekutowski refer to patients who have had an ileostomy. An ileostomy is a surgical procedure wherein the large intestine is removed and the digestive tract is shortened. After this surgery, the patient's digestive tract ends at the lower end of the small intestine. Using these cases, the contents of the small intestine can be analyzed to determine the efficiency of the digestive process no matter what kind of diet has been eaten. This research provided evidence that enzymes in food do survive during digestion and can indeed add significantly to the nutritive value of the foods ingested. Enzyme supplementation further increased the digestibility of all foods.

Does raw food stimulate enzyme secretion as much as cooked food?

Raw food does not stimulate enzyme secretion as much as cooked food. Less stomach acid is secreted when raw food is eaten. This permits food enzymes to operate for a longer period in the food enzyme section of the stomach than when the meal consists of cooked food. Consequently, more digestion is performed by food enzymes. Thus, when we eat raw food, the enzymes in the raw food lessen the excessively high amounts of digestive enzymes. Thus, the pancreas and saliva, which normally secrete these enzymes, are not required to work overtime. In addition, raw food enzymes are much less concentrated than pancreatic digestive enzymes. This means that digestion of a raw food meal takes more time.

I've heard that enzymes are not destroyed once they have been used. Is that true?

Many college textbooks teach that enzymes are not destroyed and consequently are never completely used up during their enzymatic action. However, other research, such as at Wright University, disputes this concept. Studies have found that enzymes do more work at higher temperatures than they do at cooler ones. The higher the temperature the faster they are used up. This refutes the idea that enzymes are not used up. When we acquire a fever, our body's enzymes work faster than at normal temperature. Research has found that after a fever has subsided numerous enzymes are found in the urine.

Other evidence shows us that human beings have the lowest levels of starch-digesting enzymes in their blood than any other creature. We also have the highest level of these enzymes in the urine, meaning they are being used up faster. Here again, we have evidence that enzymes wear out and are discarded.

Can our bodies make enzymes to replace our worn out and excreted ones?

The body can make enzymes. However, research[15] confirms that it is self defeating to obligate the body to make excessive amounts of highly concentrated digestive enzymes. This places an unnecessary drain on the rest of the organs and tissues. Stress and hard physical labor in hot temperatures seem to use up more enzymes which could significantly shorten your life. To prevent this enzyme loss from doing so, you have only one solution. You must provide enzyme reinforcements from an outside source. This cuts down on the need for the body to secrete digestive enzymes, and allows it to make a sufficient quantity of metabolic enzymes.

As we age, does our ability to produce enzymes decrease?

Yes. Research[21] found thirty times more enzymes in the saliva of young adults than in that of persons over sixty-nine years of age. Another researcher[22] found higher levels of amylase in the urine of young adults as compared to older adults. There is an abundance of literature that shows that rats and other experimental animals live longer when their food is significantly reduced. An explanation for this could be that less food means fewer digestive enzymes are required to digest the reduced food intake. This could contribute to a higher enzyme potential, which could delay death as well as arming the body against disease.

Why does our ability to produce enzymes decrease when we get older?

Bartos and Groh[3] enlisted ten young and ten older men for a study in which they used a drug to stimulate the pancreatic juice flow. The juice was then pumped out and tested. It was discovered that considerably less of the enzyme amylase was present in the pancreatic juices of older men. This enzyme deficiency of the older group was deemed to be the result of exhaustion of the cells of the pancreas. Other research[22] indicates that not only are there fewer enzymes in the pancreas but also in the trillions of cells in our bodies as we age. One explanation for this might be that our pancreas, which weighs only three ounces, cannot begin to supply the vast amount of enzyme activity required for pancreatic secretion. This does not even take into account the tremendous need for protein to equip the enzyme complex. The pancreas must borrow from entities stored in the cells to make the enzyme complex. Perhaps this is a definition of old age. Old age and debilitated metabolic enzyme activity are synonymous. If we postpone this debilitation, then we could very well delay the aging process.

Can supplemental enzymes slow our aging process?

Most probably. Research[20] on rats that were given supplemental enzymes showed that the supplemented rats had more enzymes than the control group of rats. Evidence clearly indicates the existence of a fixed enzyme potential in all living creatures. This potential most probably diminishes in time, subject to the conditions and stresses of life. If we eat an enzyme-free diet we most probably use up a tremendous amount of enzyme potential by over-secretions of the pancreas and other digestive organs. The result may be a shortened life span, illness and lowered resistance to stresses of all types, psychological and environmental. By eating foods with their enzymes intact and by supplementing cooked foods with enzyme capsules, we might stop abnormal and pathological aging processes. By the way, the enzyme-fed rats lived three years in comparison to two years for those rats that were fed an enzyme-free diet.

I've heard that Eskimos experience very little sickness in their culture. Why?

The Native American word **Eskimo** means "he eats it raw." The original habits and customs of the Eskimo can give valuable insights in how to achieve a high level of health. For one thing, the Eskimo practiced conservation of body enzymes by arranging for outside enzymes to help digest his food. Plant food is virtually nonexistent in the high Arctic. The Eskimo had to adapt to what was available and was forced to modify animal flesh in several ways. It served him not only as fuel but as food that maintained excellent health and prevented disease. He accomplished this by using the food enzymes, cathepsins, found in great abundance in meat and fish. The Eskimo utilized this in both predigestion and digestion. Remember, cathepsin is an enzyme found in animals that breaks down the tissue of dead animals. The Eskimo would allow his fish, seal, polar bear, and caribou to autolyze (decompose) itself before he would eat it. The food that was left after this process was full of enzymes (lipase, protease, and amylase) that kept the Eskimo from wasting his own digestive enzymes. The secret to the good health of the carnivorous Eskimo is not that he eats various meats, but that he forbids his personal enzymes to digest all of it. We can do the same with protein, carbohydrates, and fats that come from plant sources.

How do enzymes work in the stomach?

Food entering the stomach is called a bolus. The stomach has two distinct divisions—fundus (upper part) and pylorus (lower part). The bolus remains in the upper part for approximately one hour. This is where predigestion of the bolus takes place. The fundus is where digestive food enzymes begin to break down the bolus into carbohydrates, fats, and protein. Raw foods supply their own digestive enzymes, thus saving the stomach from supplying all the enzymes. Cooked foods, which have no enzymes, must wait in the fundus until the stomach supplies the enzymes. Predigestion by food enzymes occurs in every creature on earth. The only exception is the human being on an enzyme-free diet. The upper section of the stomach, the fundus, has no peristaltic (movement of food) acid or pepsin, and therefore if enzymes are absent from the diet only minimal digestion can take place.

The lower stomach (pylorus) performs the second step in digestion, but only of protein. In the lower part of the stomach, pepsin, a powerful digestive enzyme, and hydrochloric acid continue the digestive process. The predigested food now enters the small intestine where it is now referred to as chyme. Here, the pancreas and small intestine cells secrete their enzymes to further break down the chyme into glucose (carbohydrates), fatty acids (fats), and amino acids (proteins) for absorption into the villi (absorption cells in the small intestine). The human stomach is really two stomachs with separate functions. Our stomachs have been provided with the means of letting outside enzymes help with the burden of digesting food. Thus, we do not have to make all of our own enzymes to digest our food. This will allow us to make more metabolic enzymes as needed and perhaps allow us to become healthier.

Can food enzymes fight diseases?

Yes. There is a connection between the strength of our immune system and our enzyme level. The more enzymes we have, the stronger our immune system will be. Hence, we will be healthier and stronger. For example, leukocytes (white blood cells) have eight different amylase enzymes that aid them with latching onto foreign substances. These substances are reduced by the leukocytes into a form the body can eliminate. Research[20] shows that leukocytes increase after cooked food is ingested. This indicates a definite compensatory measure on the part of the body to transport more enzymes to the digestive tract for digestion. There is no increase in leukocytes after raw food is consumed. Studies suggest that enzymes are related to all diseases via the immune system, whether the disease is acute or chronic. If the pancreatic output of enzymes is hindered, the whole body is affected. Therefore, we must eat raw foods or take supplemental enzymes to fight infection and any other invaders that can weaken our state of health.

How do enzymes affect our endocrine system?

Cooked foods, devoid of enzymes, overwork the endocrine glands. This leads to diseases like hypoglycemia, diabetes, and obesity. The endocrine system, especially the thyroid gland, and the brain, more specifically the hypothalamus, work together to regulate the appetite. The glands know when we have had enough to eat, a condition known as "satiety." When this occurs, they will close down the feeding center in the brain. Eating raw food takes stress off the endocrine system. If the enzymes that usually accompany food are not present because of being overcooked, the glands will not find enough nutrients or calories. They, in turn, overstimulate the digestive organs, demanding more food than is needed to maintain our health. This results in excessive secretion of hormones which cause overeating, obesity, and finally exhaustion of the hormone-producing glands. The enzyme potential becomes even more depleted as it struggles to carry on the increased metabolic activity. Over-stimulation of the pituitary gland, known as the leader of all the endocrine glands, will cause a false feeling of well-being. Eating enzyme-free foods could cause other organs like the pituitary, the thyroid, the adrenals, and the reproductive glands (not directly involved in digestion) to become hindered as well. It is important to realize that the enzymes, the circulatory system, nervous system, endocrine system, and the digestive system are connected and interdependent in life processes.

How do enzymes affect the pituitary gland?

Cooked and refined, enzyme-free food causes a drastic deviation in pituitary gland size and appearance. Animals fed enzyme-free foods revealed damage to their pituitary glands similar to that of humans who were fed cooked foods. This damage included decreased vascularity, increased connective tissue, and a change in the distribution of cell types within the pituitary. The relationship of enzymes to the pituitary was also demonstrated when surgical removal of this gland from animals revealed that the enzyme level in the body of the animal decreased significantly. This research demonstrates that hormones influence the activity of enzymes, and enzymes are necessary in the formation of hormones.

Can enzymes control obesity?

Very definitely. Obese individuals were found to have a deficiency in the enzyme lipase. Lipase is found in abundance in raw foods. Cooking destroys the lipase in foods. Lipase is the enzyme that aids the body in the break down of fats. Without lipase, our fat stagnates and accumulates in our arteries which could lead to heart disease. Lipase also helps us to burn fat for energy. Cooked foods that have no enzymes will put weight on more abundantly than raw foods. For example, pig farmers will not feed their pigs raw potatoes because the pigs stay lean. However, when the farmers feed the pigs boiled potatoes the pigs get fat.

Another reason why enzymes reduce obesity is because cooked foods cause drastic changes in the size and appearance of the pituitary gland. Research[33] has found that enzymes affect our hormone-producing glands. Hormones influence our enzyme levels. Cooked foods cause our pancreas, thyroid, and pituitary glands to exhaust their enzyme supply in order to digest our foods. This makes the body sluggish, and weight is gained. Raw food calories are relatively non-stimulating to glands and stabilize body weight more so than cooked food calories.

Is there a significant difference between raw calories and cooked calories?

Well documented studies have proved that eating cooked starch causes the blood sugar of diabetics to increase significantly. In one study normal non-diabetic and diabetic subjects were fed raw starch and then had their blood tested for sugar. The diabetics who participated in this research found that their sugar levels rose only 6 milligrams the first half hour. Then it decreased 9 milligrams after one hour, and 14 milligrams two hours after ingestion of the raw starch. In the non-diabetic persons, there was a slight increase followed by a slight decrease in the blood sugar in one hour. This research indicates that there is a difference between raw and cooked calories.

Is raw fat as fattening as cooked fat?

From examination of the research in this area, it is hard to escape the conclusion that raw food is not fattening in the conventional sense. Isolated Eskimo tribes ate raw meat as a part of their diet. They ate raw blubber, raw butter, and other raw fats and yet there is no obesity in their society. Furthermore, Eskimos had no hypertension, no heart disease, and no hardening of the arteries. This is explained by the fact that all raw fats contain the enzyme lipase which helps digest fats properly. Lipase is absent in cooked fats.

Can you get fat eating raw foods?

You could, but it is highly unlikely. Avocados, sweet potatoes, bananas, and pears contain a fair number of calories. However, its very hard to gain weight from eating them. It would be an exceptional person who could get fat eating too many bananas. Furthermore, these foods do not clog up the arteries. However, because of the body's need for protein, I would not recommend trying to rebuild it with only fruit and vegetables.

Do obese people have a shortage of enzymes?

There is some evidence that obese individuals do have a shortage of lipase. Researchers at Tufts University School of Medicine[20] conducted some tests on the abdominal fat of eleven extra-heavy individuals (average weight of 340 pounds) and found a lipase enzyme deficiency in their fat cells. This could be explained by the fact that obesity and abnormal cholesterol deposits both have their beginnings in our failure to permit fat predigestion in the upper stomach. This occurs due to the destruction of the natural lipase content of fatty foods.

I like to eat a meal of meat, potatoes, bread, and salad. Is that bad?

It could be made better if your salad contained a lot of raw foods. However, your garden salad of raw vegetables does not contain enough enzymes nor all the right enzymes to digest meat, potatoes, and bread. The cooked meat, potatoes, and bread do give you an abundance of minerals and vitamins, but not enough enzymes. Try to adapt to a diet that contains approximately **75 percent raw calories and 25 percent cooked calories.** Eat as many fruits, vegetables, legumes and grain foods in their raw state as possible. If you cannot make yourself eat these raw foods, then you must take enzyme supplements if you desire a long and healthy life.

What are supplemental enzymes?

Supplemental enzymes are those that have been extracted in some manner from either plants or animals. They are taken in addition to a normal diet. Pepsin, an enzyme that digests proteins, was the first enzyme used by doctors to help with protein digestion. Pepsin is extracted from the stomach of pigs and requires a very low pH in order to be used by the human body. It is best used in skin products for exfoliation of the skin, or in meat tenderizers. Another enzyme supplement was made from the pancreas of slaughterhouse animals. This enzyme could digest proteins, carbohydrates, and fats. However, the pancreatic enzymes work best in an alkaline medium present in the duodenum. Pancreatic enzymes will not work in the acidic stomach and cannot perform predigestion. There is little need to take these supplemental enzymes for digestive uses.

In order for supplemental enzymes to work they must be able to help with predigestion in the upper stomach, the fundus. The Japanese have developed a method for extracting protease, amylase, cellulase, and lipase enzymes from fungi that work throughout the entire digestive system. Although there are hundreds of varieties of aspergilli, the strains used in the fermentation of plant enzymes (**Aspergillus oryzae** and **Aspergillus niger**) have tested to be free of mycotoxins (fungus-produced toxins). Extracts of these enzymes are dried into powders and put in capsules. These enzymes should be taken with a meal if predigestion is to be the most effective. If you wait until after the meal, you delay the action of the enzymes on the food.

Is it possible to treat yourself with enzyme supplements for health problems?

Treatment requires specialized knowledge and experience. In order to be successful with therapy, which usually requires massive dosages, enzyme treatments should take place under the guidance of a health-care professional. However, taking one or two capsules at a meal for predigestion usually does not require professional guidance. Rather, that is regarded as nutritional supplementation. You are replacing enzymes that are supposed to be in your food, even if they are not. Taking more capsules requires careful professional supervision, usually for prolonged periods.

I've heard that some foods contain enzyme inhibitors. Is that true?

Yes. Tree nuts, seeds, beans, and grains contain enzyme inhibitors, along with a very active number of enzymes. But because enzymes are very active entities, nature had to put reins on the seeds and nuts and make them dormant. This slowing down by nature defines an inhibitor. When the seed falls to the ground and becomes eventually covered with soil, nature deactivates it when it absorbs moisture. This is when the seed begins to germinate or sprout, forming a new seedling. The enzyme inhibitors are no longer potent once the seed is germinated; it is at this point that it is most healthy as food for humans. Eating too many nuts and seeds while the enzyme inhibitors are active can cause GI tract problems. Taking enzyme supplements can neutralize the inhibitors.

I have arthritis. Will enzyme therapy help?

Some researchers believe that rheumatoid arthritis might be a deficiency disease arising from an inability to deal adequately with protein digestion and metabolism in the small intestine. Enzymes extracted from intestinal mucosa in the small intestine were given to persons with rheumatic ailments. Among more than seven hundred patients treated with these enzymes over a period of seven years, good results were obtained in the treatment of rheumatoid arthritis, osteoarthritis, and fibrositis. In addition, cases of ankylosing spondylitis and Still's disease have responded to enzyme therapy.

It should be pointed out that for the first two or three months of enzyme therapy there may be no noticeable improvement. The longer the duration of the disease, the longer the lag before improvement is observed. However, a massive dosage may speed up the process. Massive enzyme therapy requires a doctor's supervision and blood tests to determine how many capsules can be tolerated. Each case is different and must be treated on an individual basis.

Can enzyme nutrition help me fight cancer?

It should be pointed out that cancer is a complex problem that requires trained medical personnel if enzyme therapy is to be employed. Research has shown abundant laboratory proof of profoundly disturbed enzyme chemistry in cancer. Most cancer cells show a deficiency in enzymes. In order for a normal cell to work properly it must have good proteins, vitamins, minerals, etc. (the forty-five known nutrients) reinforced daily. But to eat these nutrients in a proper diet is not enough. Metabolic enzymes, those within the cell itself, are required to transform these materials into blood, nerves, organs, and tissues. If you allot much of your enzyme power for digestion, and less for running your body, you are inviting cancer. Through outside enzyme nutrition for digestive support, you will have plenty of enzyme power to run the cells properly. This could possibly prevent or help to fight cancer.

How do supplemental enzymes remedy or prevent cancer?

Research is unclear; the jury is still out. However, there are two theories that researchers postulate as the possible answers. First, researchers can trace the beginning of cancer to the cell's DNA. Cancer cells begin with an alteration of normal DNA in the cell. Promoters such as stress, saturated fat, obesity, ultraviolet rays from too much exposure to the sun, and other factors cause the four nucleotides (adenine, guanine, cytosine, and thymine) that make up DNA to change their normal chemistry and cause the cell to make the wrong protein. Once the wrong protein is made, the cell will continue to make the wrong protein forever. These cells spread (metastasize) and destroy other cells. Probably at any given time your body has approximately one hundred to maybe ten thousand of these cells attempting to make the wrong protein. Do not worry, nature has provided repair enzymes that travel up and down the DNA structures and correct the errant proteins. If you take supplemental digestive enzymes, perhaps the body can make more of these repair enzymes to help prevent cancer cells from forming.

The second theory proclaims that cancer cells are covered with a protein sheath that protects them from the attack of white blood cells. The white cells will not attack the cancer cells because they cannot recognize them as dangerous to the body. Many researchers believe that the protein covering disguises the cancer cell, making it appear friendly to the body. Taking an enzyme supplement, especially protease, the great digester of proteins, may break down the outer protein covering of the cancer cell. This leaves it wide open to assault by the white blood cells which will recognize it as an invader and rush to destroy it.

I have several allergies. Can enzymes help me?

Many researchers theorize that being allergic to a raw food may be nature's way of telling us that the food's enzymes are incompatible with some unhealthy bodily condition. The natural function of the enzyme is to try to destroy the unhealthy condition. This confrontation between food and enzyme and disease could result in the classic symptoms of itching, nasal discharges, and rashes. There are various types of metabolic enzymes, including scavenger enzymes. Scavenger enzymes are believed to patrol the blood and dissolve the waste that accumulates from the millions of metabolic reactions that take place each second within each cell of the body. In fact, some of our scavenger enzymes are present in white blood cells. The main function of these enzymes includes the attempt to prevent the arteries from clogging up and the joints from being crammed with arthritic deposits. If the scavenger enzymes find the right substrate they latch on and reduce it to a form that the blood can use for proper disposal. If these scavenger enzymes cannot handle the waste, nature causes some of the them to be thrown out through the skin. Sometimes these wastes are thrown into the membranes of the nose and throat and produce the familiar symptoms that we call allergies.

Other researchers believe that allergies are caused by incompletely digested protein molecules. Allergies can be treated and often alleviated if certain enzymes are taken that can act as scavengers or as protein-digestive enzymes.

Can taking enzymes help prevent heart disease?

Enzymes might be effective in helping certain kinds of heart or cardiovascular disease such as atherosclerosis, hypertension, and other blood vessel diseases. There are several causes attributed to cardiovascular disease. One popular theory held by many researchers is an impairment of fatty metabolism by the body's cells. First, this faulty fat metabolism begins with incomplete digestion and absorption of fat by the intestinal cells. Second, fat metabolism is also hindered at the cellular level. Poor enzyme activity, especially lipase activity, is clearly a factor in this faulty metabolism. Lipase can come from three different sources. First, lipase can be present in the food we eat. There is an abundance of lipase in raw butter and raw animal fat. Although Eskimos eat both of these foods, cardiovascular disease is virtually nonexistent in their society. Second, supplemental lipase works in the fundus part of the stomach which helps break fat down in the predigestion stage. It is here that lipase breaks the fat down into the substrate form that eventually makes it easier for the lipase produced by the pancreas to break it down even further. The third source of lipase can be found in the duodenum; the upper part of the small intestine. This lipase is called pancreatic lipase. Some researchers theorize that it is possible that when lipase from an outside source works on fat in the predigestion stage in the stomach, it induces certain changes. These changes enable pancreatic lipase to create a better finished product than when it must do the entire job on its own. Thus the pancreatic lipase can break the fat down to its most desirable and absorbable form if supplemental enzymes are taken with the meal.

Can enzymes lower cholesterol?

Yes, in most individuals. Remember, cholesterol is a form of fat. Research substantiates that consumed animal fats tend to cause cholesterol to settle in the arteries, thereby allowing the onset of atherosclerosis. However, it has also been found that the crystal-clear "purified" vegetable oils, if not heated, do not raise the blood cholesterol level. One answer for this might be that lipase is found in these unheated vegetable oils. One study[14] found that the fat tissue in obese humans has less lipase than that of a slender person. This could indicate the need for supplemental enzymes.

Three British researchers[17] tested the enzymes in individuals with atherosclerosis to find the relationship between cholesterol and clogged arteries. They discovered that all enzymes studied became progressively weaker in the arteries as persons aged and as the hardening became more severe. These scientists believe that a shortage of enzymes is part of a mechanism that allows cholesterol deposits to accumulate in the inner part of the arterial walls (intima). Another researcher[14] found a progressive decline in lipase in the blood of atherosclerotic patients with advancing middle to old age. Yet another study[24] showed that not only was lipase low in older persons, but that older atherosclerotic persons had slow fat absorption from the intestine. In addition, some absorbed fat was in the unhydrolyzed state. When the older and younger subjects were fed lipase extracted from an animal pancreas there was a definite improvement in fat utilization. Think of the advantage of taking plant enzyme lipase at the beginning of the meal to help in the predigestion of that food.

Can enzymes help with diabetes?

This depends on what type of diabetes you have. Type II (adult onset) normally responds better to enzyme therapy than Type I (juvenile). Research has shown that when there is a lack of blood amylase, blood sugar levels can be higher than normal. When the enzyme amylase is administered, the levels drop significantly. One researcher[31] showed that 86 percent of the diabetics whom he examined had a deficiency of amylase in their intestinal secretions. He administered amylase to a majority of these patients, and found that 50 percent of those diabetics who were insulin users could control their blood sugar levels without the use of insulin. Amylase may help with storage and utilization of sugar in the blood. Another study[25] found that cooked starch foods, in which amylase and other enzymes are destroyed, caused the blood sugar levels to rise significantly in one half hour after ingestion. After two hours the cooked-food-starch eaters' blood sugar levels fell quickly and significantly. This resulted in fatigue, anxiety, and sluggishness. By comparison, the raw-starch eaters' blood sugar levels only showed slight rises and drops. These patients experienced much more steady metabolic rates and emotional stability. Many diabetics could lower their insulin requirements if they would eat raw foods and take enzyme supplements.

Can enzymes help with hypoglycemia?

Authorities have estimated that anywhere from ten to one hundred million Americans are suffering from hypoglycemia. If we are one of those millions, every organ in our body is going to be affected by the low blood sugar level. A drop in blood sugar level will cause mental fatigue, depression, and sluggishness because our brain depends on glucose for its food. Our endocrine glands, especially the pituitary, adrenals, thyroid, and pancreas control our glucose (sugar) level. The pancreas secretes insulin that causes a decrease in our glucose level by facilitating the movement of glucose into the cell. Glucagon, another pancreatic hormone, raises our glucose level when it drops too low. Our thyroid glands secrete thyroxin, which controls the rate of our cells' basal metabolic rates and usage of oxygen for energy. All these glands are controlled by the pituitary gland, which in turn is controlled by an area of the brain called the hypothalamus. The hypothalamus receives information from all parts of the body via the nervous system. These signals convey everything from hunger, depression, and happiness to exhaustion and sluggishness.

When there is a deficiency of enzymes from our food the pituitary and other organs can hypertrophy (enlarge). If this happens, we are more susceptible to disease, particularly hypoglycemia. Taking amylase, either through supplements or by eating raw foods, causes the glucose level to stabilize, thus protecting against an erratic rise or drop in the blood glucose level. This helps alleviate depression, fatigue, and sluggishness.

Should children take enzymes?

Yes. Children usually eat the same enzyme-deficient foods as their parents. The importance of breast feeding should be emphasized as a way for an infant to acquire enzymes. Children who are breast fed acquire dozens of enzymes from their mother's milk. Bottle-fed babies receive pasteurized milk that has been heated, which destroys the milk enzymes. This causes the baby's enzyme factory to begin using its enzyme potential from the moment of birth. Research[31] indicates that this could be harmful for the child. A study involving 20,061 babies divided them into three groups: Breast fed, partially breast-fed, and bottle fed. These physicians studied the morbidity (sickness) rate for the first nine months of the infants' lives. They found that 37.4 percent of the breast-fed babies experienced sickness in comparison to 53.8 percent for the partially breast-fed and 63.6 percent for the bottle-fed infants. It is obvious that babies who were entirely breast-fed had far less sickness than babies who were only partially breast-fed or who were bottle-fed. They studied the mortality rates of these three groups as well. The results follow.

	Number of Infants	Total Deaths	% of Deaths
Breast-fed	9,749	15	0.15
Partially breast-fed	8,605	59	0.7
Bottle-fed	1,707	144	8.4

The mortality rate among the bottle-fed infants was fifty-six times greater than among the breast fed.

Dr. Anders Hakannson at Lund University in Sweden discovered that when he added mother's milk to cultured cancer cells that were alive prior to the addition of the milk, soon thereafter the cells died. Further tests indicated that only tumor cells were killed by the milk, while normal adult cells were left intact.

Research findings certainly seem to be telling us that we, including pregnant women and children, must eat raw foods that contain enzymes or take plant enzyme supplements. Or perhaps we are supposed to be doing both.

Can enzymes help fight leukemia?

Leukocytes are produced in greater quantity by our bodies when we eat cooked foods. Leukemia is a disease in which the body produces too many leukocytes (white blood cells). Leukocytes are rich in enzymes and are transported to the stomach area to aid in digestion when we eat food devoid of enzymes. Eating raw food will help to alleviate the condition by eliminating the excess of leukocytes produced when cooked foods are ingested.

Many people with leukemia have been successfully treated with raw food diets. Research[20] has demonstrated that: (1) Raw food produced no increase in white blood cells; (2) common cooked food caused leukocytosis, or an increase in white blood cell production; (3) pressure cooked food caused greater leukocytosis than non-pressure cooked food; and (4) manufactured foods such as wine, vinegar, and white sugar along with cooked, smoked, and salted meat produced the most white blood cells. Leukemia patients would be wise to eat raw foods and take enzyme supplements.

It has been the experience of many doctors that high dosages of protease enzymes administered orally to their patients will result in a balance of the red and white blood cell count.

Does the addition of raw food to the diet or juicing guarantee enough enzymes to meet our needs?

Raw foods provide only enough enzymes to digest the food containing them, in whatever quantity of that food is ingested. There are no extra enzymes in raw food to digest cooked or processed food that may be eaten at the same meal as the raw food. Because of the risk of bacterial contamination, many foods should not be eaten raw. These include meats, poultry, eggs, and beans. The fiber content normally found in raw food is difficult to digest.

Could you comment on how enzymes affect our colon?

Research[9] indicates that cooked food passes through the digestive system more slowly than raw food. In the small intestine and colon any partially digested food ferments, rots, and putrefies. These harmful substances are absorbed by the body only to cause gas, heartburn, and terrible degenerative diseases like arthritis, cardiovascular conditions, cancer, and diabetes.

Colon researchers have found that much of the body weight can be just waste accumulated within the one hundred thousand miles of blood vessels, the lymphatic system, bone joints, and intra-extra-cellular regions. The largest amount of waste is found in the impactions within the colon structure. This waste can add up to fifty pounds of body weight if the diet consists of only cooked or greasy foods. Unfortunately, some of this partially digested cooked food found lining the small intestine and colon passes into the bloodstream and is deposited as waste throughout the system. If this waste is composed of calories, it manifests as obesity. As excess minerals, it creates arthritis. Excess protein waste can become cancer, fat leads to high cholesterol, and sugar will eventually cause diabetes. Enzymes in the diet will digest the food into smaller particles more easily assimilated by the body. This will greatly help to keep this impacted fecal matter from remaining in the colon.

I have migraine headaches. Will enzymes help?

Some migraine headaches are caused by hormonal imbalances or a dysfunction in one of the endocrine glands. The pituitary gland controls all the other endocrine glands by influencing their hormone output. Eating enzyme-free foods will overstimulate the pituitary gland. This may cause excess secretions that could lead to a hormonal imbalance. Other migraine headaches could be caused from the toxins in the colon produced by poorly digested food. Eating raw foods or taking certain enzyme supplements has been shown to be beneficial to many sufferers of migraines.

I have insomnia. Will enzymes help?

There are many causes of insomnia. Causes associated with hormonal imbalances in the endocrine system have responded favorably to enzyme therapy. The lack of metabolic enzymes will definitely affect the secretions of the pituitary gland, which could lead to insomnia. Eating raw foods and taking plant enzyme supplements will enhance your endocrine system.

Can enzymes help me with my psoriasis?

Many dermatologists have reported favorable results with enzyme therapy. In the early 1930s researchers treated their patients by having them eat large quantities of raw butter. Raw butter contains high levels of lipase. Recent research has determined that massive doses of lipase will help cure psoriasis. However, when large quantities of concentrated enzymes are prescribed, it is essential that the patient take them under the guidance of a qualified health-care professional.

Are plant enzymes capable of surviving the stomach's hydrochloric acid better than animal enzymes?

Research supports the theory that plant enzymes are much more capable than animal enzymes of surviving the harsh acidic environment of the stomach. Dr. Selle[20], a physiologist from the University of Texas, fed dogs cereal starch with the addition of the starch-digestive enzyme amylase, extracted from barley. The pancreases of the dogs were tied off so that no pancreatic digestive juices would affect digestion. The stomachs were emptied after a certain period of time and in some cases, the cereal starch was 65 percent digested. The barley amylase was proven to digest the starch in the stomach with a pH as acidic as 2.5. Furthermore, Dr. Selle's research revealed that one half hour after amylase was taken orally, 71percent of it was found in the small intestine still actively digesting starch. When the fecal matter was analyzed for its enzyme content to determine if any of the plant amylase was still present, there was a higher level of it than the body's pancreatic enzymes.

Enzymes that are released from the pancreas can only function in the alkaline conditions that are found in the duodenum. Usually animal enzymes are extracted, purified, and concentrated from pancreas of slaughterhouse animals. As a supplement, they cannot function in the predigestive stomach where much of the digestion takes place. As Dr. Selle found in his study, plant enzymes digest food in the stomach as well as in the small intestine. This relieves the pancreas from having to secrete its enzymes to digest the food.

Do athletes and physically active people have different enzyme needs than others?

Yes, in theory athletes have a greater need for enzymes. Research shows that enzymes are lost in perspiration. In addition, the body uses up an amazing quantity of enzymes during exercise. This is especially true for those who push their bodies past their endurance levels. This is evident in the cramping and dehydration problems experienced by many people who work out, run or overdo their exercise routines.

I'm trying to "bodybuild" and I want to know if enzymes will help me with this?

Most likely. If the naturally occurring enzymes are not working in the muscle tissue, there would be no muscular growth. There would not be even the basic muscular activity necessary for growth. Enzymes are the catalyst that turns food into energy, thereby allowing the muscles to move and grow.

Appendix I

Seven Stages of Disease Progression

Appendix I

Seven Stages of Disease Progression

It has been my experience to witness fear, surprise, and utter confusion by a person receiving a negative diagnosis. Disease is never supposed to happen to us. We believe that this condition happened to us overnight and there was absolutely nothing that could have been done to spare us. Many of us feel like victims and completely without choice in this matter. I have been in this position more than once in my lifetime and understand the fear and feeling of helplessness. However, what I have since learned is there are *choices,* and most conditions do not happen overnight. The choices a person makes not to ever exercise, or to stop for lunch every day and eat fast foods, or lead a stressful, negative lifestyle that can lead to the depletion and an imbalance of enzymes. Every one of us is born with an inherited ability to make metabolic enzymes. However, enzymes can either be repressed or destroyed when the body's condition or environment is not in balance.

It is these poor *choices* that, in fact, result in the imbalances to the whole body that can allow disease to happen. No, disease does not happen overnight. It takes years of this constant neglect and over-indulgence to alter our enzyme balance and create a disease condition.

STAGE ONE: *This is characterized by approximately 95% lifestyle and/or environmental factors.*

The first stage of disease can be caused by over-eating, drinking, overwork, worry and over- indulgence in cooked and processed food. Foods that are cooked or processed (unlike raw foods) lack the enzymes to assist in their digestion. Stress and alcohol deplete valuable enzymes. Sugar, salt, caffeine, soft drinks, tobacco, drugs, serum injections, impure water, polluted air, tension, depression, and lack of rest only add to the extra burden for the body to produce the necessary enzymes required for good health. The result is a weakened immune system and reduced vital energy or strength. How many of us *choose* to do this on a regular basis and never stop to think about the outcome?

Our metabolic enzymes (bio-chemically produced in the body) have now become compromised. Even the simplest efforts of the body (like mastication) to receive the proper enzyme support from our raw food is now void. If we could aid the body by adding digestive enzyme supplements such as protease to help break down the proteins, amy1ase to break down the carbohydrates and starch, and lipase to break down the fats, we have at least begun to help the body retain its balance.

STAGE TWO: *The choices above when not corrected appropriately, will progressively overwhelm the system and create this stage.*

The second stage is created when the body's ability has become so compromised, it cannot even digest food. Result—there is an increase in food toxins. The initial lack of enzymes, created by the first stage, have slowed down the elimination of toxins by the liver and kidneys and has placed a burden on the body's defense system. Poor nutrition from stage one did not provide the necessary elements to supply the body's building blocks for metabolic enzymes, antibodies, and even hormones, and free radicals have taken their toll. These factors can be accentuated with age, as the body bio-accumulates various toxins which provides a continuous weakening.

During digestion, a lot of blood must flow to the digestive organs. Energies of the organism are always divided between assimilation and elimination. An unequal use of those energies is again caused by the practice of overeating, and the other abusive choices mentioned. Science tells us that we use 80 to 90 percent of our body's energies just digesting food alone. Other stimulating choices also divert energy. The stimulating foods as an example are caffeine, sodas, sugars, and alcohol. When non-usable substances and their toxins are not or cannot be eliminated, they are stored in the tissues, joints and organs. This leaves the body intoxicated, sluggish and weakened. One of the recognizable signs may be constipation, but there are other signs of toxicity that begin here. Now the body is not only overburdened with toxins, but the struggle of the body's enzymatic potential to keep up the proper digestive enzymes, hormones, rebuild muscle, tissue, bone, etc. is now impaired. A sluggish, continually tired body is susceptible to crises.

Simple over the counter digestive aids do not work here. It takes a natural supplement source to replenish a proper digestion to the body. Other chemically based substances will only add toxicity to the body. The enzymes mentioned in stage one need to be increased.

STAGE THREE: *Progression to stage three involves the pH imbalances and a definite deterioration to the biological terrain. Cellular activity is compromised.*

The third stage of disease progression is the acid/alkaline imbalance that must then follow. When continual poor eating and drinking habits that lead to an over toxic body are consistently repeated, then an over- acid body fluid condition develops called acidosis. Acidosis means the body's alkaline reserves are now being depleted. Alkaline buffers in the blood are the body's defense against acid-forming foods, toxic waste, stress, etc. Alkaline mineral elements, when not available in the bloodstream, are leached from teeth, bones, liver, etc., to neutralize deadly acids and preserve life. When there are not enough buffers for balance, these acids enter the cell and eventually cause the cell itself to become saturated by acid. Once our cells are saturated, we will experience a loss of electrons (energy) from the cell itself. Potassium, chlorides, magnesium and other minerals, along with our life's energy, will then be carried out of the body via the kidneys.

This acid/alkaline imbalance further impairs our digestive ability because the body cannot produce hydrochloric acid in the stomach for digestion. The absence of free hydrochloric acid in the stomach leads to chronic gastritis.

The burping, belching and gas that can result from these chronic problems can again be alleviated by the use of supplemental digestive enzymes. Over the counter antacids only increase this problem, because remember, the body's number one priority is to keep the system in balance. If we add alkaline, the body will create acid to sustain its balance and the cycle becomes repeated over and over.

STAGE FOUR: *The continuance of constant toxicity have created a depressed environment in the body and the symptoms increase.*

This stage coincides with further deterioration of the metabolic environment (presence of high levels of toxins). Various parasitic and pathogenic factors will find the conditions favorable for invading the body. In fact, there is an accumulation of metabolic wastes which includes dead cells. There is an inability of the waste removal system to adjust to the load imposed on it. There is metabolism and production of toxins from the unwanted micro-flora.

Accumulation of toxic materials within the organs, joints or other vital parts of the body become apparent. The toxicity becomes evident. For example, of the 25 trillion red blood cells alone, seven million die per second and their by-products contain toxins such as carbon dioxide. Anabolism is greatly affected and emphasis of the body's cellular activity is directed to defense. Additionally, the buffering capacity of the body fluids is impaired. [We refer to this stage as inflammation.]

Series of symptoms manifested by stone-like formations in the joints, kidneys and gall bladder, which consist of trapped inorganic substances from poorly digested food, drink, drugs may be noted in some people. These toxic substances are indigestible, but at the same time, cannot be expelled by the weakened body.

For example, so-called "allergies" are nothing more than the irritation of already chronically inflamed nasal passages from a pronounced toxemic condition. Toxemia is the basic cause of all inflammation, including the lining of the membranes of the hollow organs of the body. Heat, redness, swelling and pain characterize this stage. Science has now proven that all allergies, whether airborne or food, begins in the gut.

With the body's immune state being compromised, protease now becomes a vital supplement for the body to survive these attacks. Protease not only becomes critical to the digestion of important proteins, but will also bind to alpha 2-macroglobulin and be delivered to the sites of immune function. Years of clinical experience has shown that toxins are also removed from the blood, perhaps from an over-all improvement in blood flow.

STAGE FIVE: *The on-set of the actual disease due to organ deterioration and functional impairment.*

The fifth stage of degeneration is characterized by further weakening of the body, frequent recurrence of ailments, and "disease" on-set. This is when a gradual thickening (hardening) of the mucous and sub-mucous tissue takes place. This is due to the causes of inflammation, which remain unchanged. This increased level of hardening chokes arterial circulation (hardening of the arteries), cutting off oxygen and the food supply. The tissues clump and break down, giving rise to skin disorders. These skin problems can be skin eruptions such as moles, liver spots, callused feet, and other dead cell accumulations. The lymphatic system becomes stagnate, is unable to utilize antioxidants, and puts the body in oxidative stress and premature tissue aging. At this point, any metabolic digestive enzymes the body is to supply for digestion is completely limited by the condition of the body. Many suffer from indigestion, bloating, gas and burning sensations in the gastrointestinal area.

Supplemental digestive enzymes with each meal have increased in dosage again and become even more crucial. In addition, protease needs to be taken between meals so it can be absorbed directly into the bloodstream, and should be taken a minimum of three times a day.

STAGE SIX: *The ability of the body has now given way to degenerative diseases and major system dysfunction occur.*

This is a stage where degenerative diseases and major system function occur. For instance, instead of normal toxin elimination, excretion of toxic-laden mucous erupts from open wounds, boils, fistulas, internal ulcers, chancre (as in AIDS) and herpes simplex. These excretions can occur from the eyes, ears, nose, vagina or other body cavities. Other symptoms in this stage could be high or low blood sugar and/or anorexia, which are nothing more than SHS (sugar handling stress) due to adrenal exhaustion from a highly toxic condition. Irritable bowel syndrome and ulcerated states are now seen in this stage. The good and bad bacteria in the colon are now out of balance and the bad bacteria are striving to take over the enzyme sights of the villi in our intestines. The overgrowth of the bad bacteria fight for control of the same sights so they can eat off of the poorly digested food sources. The ability of the body to breakdown wheat or other grains at this stage is further handicapped by the overgrowth of bad bacteria.

The conditions will worsen. Indigestion/malabsorption will favor the overgrowth of undesirable and opportunistic micro-flora in the GI tract which will convert the poorly digested foods into toxic compounds, many of them shown to induce cancer. For instance, studies have shown that excess cholesterol may be converted by some undesirable bacteria of the GI micro-flora into estrogen like compounds that are cancer promoting: thus in addition to any normal estrogen produced by the body as normal physiological process, more estrogen may be produced by poor digestion creating an excess imbalance in the body.

Probiotic supplementation ("friendly" bacteria) now has become not just a prevention, but a necessity for the health of the whole. A preferred probiotic is the lactobacillus plantarum (l. plantarum) with the ability to eliminate or reduce most other bad bacteria and fungi as it is the dominating habitat of all sorts of naturally fermented foods. L. plantarum has an upregulatory effect on the immune system, changes the immune cells and influences production of cytokines. One consequence of this is the normalization of the colon pH.

STAGE SEVEN: *The body is now being overwhelmed by disease.*

In this final stage there is a cellular disorder that causes tissue growth of a morbid nature. Poor circulation and general degeneration of the body resulting from continued abuse hasten this growth. It is in this stage in which the cells can no longer renew themselves adequately. Tumors and other types of growths are allowed to grow.

Cancer is formed as a result of physical, chemical, or genetic mechanisms that interfere with the production of the body's cells. These cells become damaged or abnormal. Abnormal cells rob the other cells of nutrients.

Healthy individuals could have anywhere from 100 to 10,000 cancer-like cells. This does not mean you have cancer because your body sends out signals for help and a group of suitable enzymes, immune cells, and cytokines react and disintegrate the cancer like cell. The toxic debris are broken down by various enzymes, resulting in their removal by the body. New cancer like cells are produced constantly but if we have a healthy and balanced immune system they will be recognized and destroyed. However, if poor eating habits, drugs that weaken the immune system, smoking, and stress upset the balance, the body's defense system breaks down as described above. The previous conditions of stages one to six have seriously compromised the body's ability to synthesize the needed metabolites, including metabolic enzymes and healthy cells.

Cancer cells are typically characterized by (Lehninger, 1982):

- **uncontrolled division of the cell and with no differentiation pattern,**
- **interference with physiological and biochemical activities of normal neighboring cells,**
- **deviation in the energy producing metabolic system:**
- **oxygen consumption is lower than in normal cells,**
- **abnormally high use of glucose than normal cells,**
- **production of high lactic acid in the cell,**
- **drainage of liver function as the high lactic acid (lactate) is converted by liver to glucose. This process makes cancer cells which are termed as "metabolic parasites." The liver has to use more of its ATP (to make the conversion) than the cancer cell generates, resulting in the weakening of the liver.**

Additionally, cancer cells may be covered with a protein coat that help them be disguised and appear as normal cells ("self") thus helping them to evade the immune system.

Damaged or abnormal cells are now left alone to coat themselves in glue like fibers about 15 times the thickness of normal cells. They hide under this protective glue-like coating so that they cannot be recognized. These sticky cells wander around looking for a place to adhere so that they can build more fibrin and multiply forming a tumor. The fiber around the cancer cell is made up of protein. Cancer cells are clever and have a "trade mark" or cell markers. Even when the cell is destroyed, the trademark can get left behind. They are so smart that they can even leave a false trail so they can be undisturbed and can form new markers elsewhere. As more and more cancer cells with specific antigens invade the body, they overwhelm the immune system.

The more cancer cells the more enzymes are needed by the body. Digestive enzyme supplements may help in meeting the body's nutrient needs, specially in cases of cancer where metabolic needs are increased. Additionally, various studies have shown that proteases ingested orally help regulate cytokines and play a role in some forms of cancer.

In Conclusion

Disease is not "bound to happen" and can be overcome if there is a willingness to eliminate all of the aforementioned abuses and maintain a disciplined adherence to a diet of life giving foods. The effect on the conditions by such a regime (digestive enzyme formulations, proper foods, and permanent life changing habits) depends largely upon the extent of degeneration of affected vital organs. If sufficient life-energy remains, the chance for rejuvenation is great enough to warrant a change. This is an example of the energy power within. Remember that all the symptoms just mentioned are nothing more than the body's effort to return to balance.

Appendix II

Systemic Enzyme Therapy

Appendix II

PROTEASES

Protease refers to a group of enzymes whose catalytic function is to hydrolyze (breakdown) peptide bonds of proteins. They are also called proteolytic enzymes or proteinases. Proteases vary in the susceptibility of the peptide bonds that they hydrolyze. Examples of proteases include: fungal protease, pepsin, trypsin, chymotrypsin, papain, bromelain, subtilisin, etc . . .

Proteolytic enzymes are very important in digestion as they breakdown the protein foods to liberate the amino acids needed by the body. Additionally, proteolytic enzymes have been used for centuries various forms of therapy. Their use in medicine is gaining more and more attention as several clinical studies are indicating their benefits in oncology, inflammatory conditions, blood rheology, and immune regulation.

Contrary to old beliefs, several studies have shown that orally ingested enzymes can by-pass the conditions of the GI tract and be absorbed into the blood stream while still maintaining their enzymatic activity. While in the blood, proteinases are taken up by alpha-macroglobulin and other anti-proteinases. However, the binding of alpha-macroglobulin to proteinases does not inactivate the enzyme, but rather ensure its clearance from the circulation after it has performed its enzymatic activity.

As in the cases of lipases and amylases, proteases are also commercially produced in highly controlled aseptic conditions food supplementation and systemic enzyme therapy. The organisms most often used are *Aspergillus niger* and *oryzae.*

PROTEASE AS SCAVENGER OF OXIDIZED AND DAMAGED PROTEINS

Oxidative reactions generate free radical damages to various molecules including proteins. Free radicals have been implicated in accelerating the aging process as well as several diseases, including diabetes, atherosclerosis, and neurodegenerative conditions[1]. Under proper conditions of nutrition and adequate activity of antioxidant enzymes, the body handles and corrects the free radical damages. However, in many instances, the body is overwhelmed by the load of pro-oxidants (free radical generating molecules), resulting in oxidative stress conditions.

One consequence of oxidative stress is the formation of oxidized proteins. Oxidized proteins often loose their function (become inactive), and undergo unfolding or conformational change of their structure which enhances their susceptibility to proteinases[1,2]. For instance, oxidized proteins in blood or extracellular fluid, include hormones, immune system proteins, transport proteins, and other proteins needed at various cellular locations.

As these oxidized proteins loose their biological function[1,2], they may not carry out the cellular tasks and biochemical reactions they are meant to perform. For instance, an oxidized hormone may not be able to attach to its receptor on the cell surface; an oxidized enzyme may not perform its activity; an oxidized antibody molecule will not bind adequately to its antigen.

Oxidative reactions occur in chains and in a cascade manner. Therefore, oxidation of one protein may lead to further oxidation reactions within the same molecule and/or other molecules which amplify the damaging effect. Thus, any oxidation of a protein if not corrected may result in impairment of biochemical functions of vital importance to the cellular viability. In order to avoid the cascade effect, oxidized proteins may be removed by an antioxidant enzyme or vitamin, or by proteolysis. Oral proteases taken on an empty stomach have been shown to be absorbed and carried into the blood stream[3,4] where they are bound to a2-macroglobulin[5].The binding of the a2-macroglobulin to proteases does not inactivate the proteolytic activity of the protease[6]. However, the complexing of the a2-macroglobulin ensures the clearance of the protease from the organism[7]. Several studies have indicated that oral proteases bound to the macroglobulins hydrolyze immune complexes, proteinaceous debris, damaged proteins, and acute phase plasma proteins in the blood stream[8,2]. It is suggested that oral proteases may help hydrolyze and remove extracellular proteins damaged by free radicals. This is based on the absorbability of the protease into the circulatory system, their hydrolytic activity and ability to remove proteinaceous debris in blood and extracellular fluid, and their susceptibility due to their unfolding and other conformational modifications from their native state.

Heavy metals and oral fungal protease

Heavy metals, such as lead (Pb) and mercury (Hg), exert their poisoning effect by binding to ionizable or sulfhydryl groups of proteins, including vital enzymes. Once they bind to an essential functional protein, such as an enzyme, they denature and/or inhibit it[9]. This interaction of heavy metals to proteins can lead to degenerating diseases, nerve damage or even death.

It should be noted that protease when taken on an empty stomach is readily taken up into the mucosa cells of the intestine and passed into the blood circulation[3,4]. Clinical observations (manuscript in preparation) have noted that upon high intake of oral protease, heavy metal concentrations have been significantly decreased in the blood. This may be due to the binding of these toxic substances with the supplemental protease enzymes, as in the use of raw egg or milk in cases of accidental mercury ingestion. This binding facilitates the removal of toxic substances, thus avoiding a life-threatening situation of poisoning. Following that same concept, the addition of a strong, wide range pH protease can spare other vital proteins from heavy metal poisoning.

LIPASE

Lipase is an enzyme that hydrolyzes the ester bonds in mono, di, and triglycerides to form fatty acids and glycerol. Although some lipase is found in the stomach, most of the digestive lipase in humans is produced by the pancreas and secreted into the duodenum where it hydrolyzes the ingested dietary fats that have been emulsified by the bile. Besides the digestive lipase in the GI tract, there is also a hormone sensitive lipase that serves to mobilize and hydrolyze lipids in adipose tissues for energy purposes in the body.

This hormone sensitive lipase is under the influence of several hormones. For instance, insulin inhibits this lipase: when carbohydrates, *i.e.*, glucose, is high such as after a meal, insulin inhibits the further release of fatty acids, thus enhancing lipogenesis, the formation of adipose tissue, as opposed to breaking down fats. However, there are other hormones that enhance the activity of the lipase when the body needs more energy and the carbohydrate levels don't need the immediate energy needs. Such hormones that release free fatty acids in the plasma for tissues to use as energy sources are epinephrine, norepinephrine, glucagon, adrenocorticotropic hormone (ACTH), alpha and beta melanocyte stimulating hormones and growth hormone.

As amylase, lipase is also produced on large scale by microorganisms including *Aspergillus oryzae*. This fungus produces a very potent lipase that is used in food supplementation.

Lipase in food digestion is very important to ensure breakdown of fats, and adequate supply of fat soluble vitamins. It should be noted that all cell membranes and other structures are made up of lipids: thus, an adequate supply of essential fatty acids in the diet are important to ensure viable cells.

AMYLASE

Amylase is an enzyme that hydrolyzes starch to disaccharide maltose and dextrins. The maltose is further broken down by maltase into glucose. The dextrins are further broken down by amylase and glucoamylase.

Amylase specifically hydrolyzes the 1, 4-alpha-glucosidic linkages in starch and similar carbohydrates. Amylase is produced by the salivary glands, the pancreas, and also by some microorganisms. *Aspergillus oryzae* and *niger* have been used in the production of commercial amylases for food supplementation.

Glucoamylase removes the D-glucose residues by hydrolyzing the 1, 4-alpha-glucosodic bonds. A particular isoform of glucoamylase in *Aspergillus niger* has been shown to hydrolyze the 1, 6-alpha-glucosidic bonds of the branching chains found in amylopectin.

NEURODEGENERATIVE DISEASES - ALZHEIMER DISEASE

Oxidative reactions generate free radical damages to various molecules including proteins. Free radicals have been implicated in accelerating the aging process as well as several diseases including neurodegenerative conditions. One of the neurodegenerative diseases shown to be caused or enhanced by oxidative stress is Alzheimer Disease. The impact of the oxidative stress on the nerve cells appears to be arising from defects in the cell energy metabolism, increase oxidation of vital proteins, modification of cytoskeletal proteins. Some of the oxidative reactions in the molecules generate covalent bonds and protein conformational changes, and glycation of proteins that may impair protein functions.

Alzheimer Disease is an age-related disorder which symptoms include progressive failure of memory, perception, orientation, and reasoning. On the organic level, the brain is the target organ and is progressively characterized by brain atrophy, nerve cell death, and various histopathological conditions resulting from neurofibrillar tangles and senile plaques made up of amyloid protein and inorganic aluminosilicate. The severity of the disease seems to be highly correlated with the presence of histopathological lesions.

Besides these histopathological conditions, Alzheimer Disease patients also have higher LDL and lower HDL when compared to control patients serum with no Alzheimer Disease. Additionally, Alzheimer Disease patients experience an active inflammatory activity. It is actually shown that in cases of Alzheimer Disease, the response of the brain to the plaques and neurofibrillary tangles is more important to the clinical manifestations than does the mere presence of plaques and neurofibrillar tangles.

According to a MIT research group, the accumulation of the amyloid precursor protein is a major factortor of the disease pathology. In cases where this precursor is degraded and not allowed to form amyloid, the disease may be prevented, whereas its conversion into amyloid leads to the Alzheimer Disease onset.

Several studies conducted conclude that any attempt to curve oxidative stress in the cells, and remove the amyloid precursor protein, called amyloidogenic, such as by proteolysis, may alleviate the disease conditions, and/or retard its onset in families genetically predisposed to Alzheimer Disease. Recent studies endorsed by the AMA indicated that the herb gingko biloba and possibly other anti-oxidants helped stop and/or slow down the symptomatic progression of Alzheimer Disease.

Additionally, proteases have been shown to be implicated in the removal of oxidized and damaged proteins, and probably the amyloid precursor protein. The removal of oxidized proteins (proteins damaged by free radicals and no longer functional) by proteolysis helps prevent further oxidation and enhance the body's ability to regulate its functions as well as the immune system.

References cited:

1. Dean, RJ, Shanlin, FU, Stocker, R., and Davies, M. "Biochemistry and pathology of radical-mediated protein oxidation." ***Biochem. J.*** **1997; 324: 1–18**

2. Grune, T., Reinheckel, T., Davies, KJA. "Degradation of oxidized proteins in mammalian cells." ***FASEB J.*****, 1997; 11: 526–534**

3. "Absorption of orally administered enzymes." Gardner MLG, and KJ Steffens, eds. Springer- Verlag, Berlin (1995).

4. Castell JV, Friedrich, G., Kuhn, CS., and Poppe, GE "Intestinal absorption of undegraded proteins in men: presence of bromelain in plasma after oral intake." *Am J. Physiolo* 1997; 273: G139–G146
5. Barrett, AJ, and Starkey, PN, "The interaction of a2-macroglobulin with proteinases." *Biochemical Journal* 1973; 133: 709–724
6. Lehmann, PV. "Immunomodulation by proteolytic enzymes." Nephrol. Dial. splant 1996; 11: 953–955
7. Salvatore V. Pizzo. "Receptor recognition of the plasma proteinase inhibitor a2acroglobulin." Pages 242–246. In: "*ISI Atlas of Science: Biochemistry*, 1988, vol. 1. Published by Institute for Scientific Information.
8. Trevanil, A., Andonegui, GA, Isturiz, MA, *et al.* "Effect of proteolytic enzymes on neutrophil FcyRII activity." *Immunology* 1994; 82: 632–637
9. *Chemistry and Life*. Hill, JW, and Feigl, DM. eds. Publisher: Burgess Publish. Company, Minneapolis, Mn. 1983. Pages 576–577

References

References for Chapters One through Ten

Beazell, J. J., Ph.D., M.D. "A re-examination of the role of the stomach in the digestion of carbohydrate and protein." *American Journal of Psychology* 132: 42–50 (1941).

Bienenstock, John, M.D. "Mucosal Barrier Functions." *Nutrition Reviews* 42:105–8 (1984).

Bockman, Dale E.; Winborn, William B. "Light and electron microscopy of intestinal ferritin absorption. Observations in sensitized and non-sensitized hamsters (*Mesocricetus auratus*)." *Anatomical Record* 155:603–9 (1966).

Bogstrom, Georg, Ph.D. *Principles of Food Science: Food Microbiology and Biochemistry.* (New York: The Macmillan Company, 1968).

Bogstrom, Georg, Ph.D. *Principles of Food Science: Food Technology.* (New York: The Macmillan Company, 1968).

Dixon, Malcolm; Webb, Edwin C. *Enzymes, 3rd edition* (New York: Academic Press, 1979).

Dutta, S.K.; Rubin J.; Harvey, J. "Comparative evaluation of the therapeutic efficacy of a pH-sensitive enteric coated pancreatic enzyme preparation with conventional pancreatic enzyme therapy in the treatment of exocrine pancreatic insufficiency." *Gastroenterology* 84:476–82 (1983).

Flatt, J. "Dietary Fat, Carbohydrate Balance, Weight Maintenance: Effects of Exercise." *American Journal of Clinical Nutrition* 45:296, (1987).

Gardner, M.L.G. "Gastrointestinal absorption of intact proteins." *Annual Review of Nutrition* 8:329–50 (1988).

Godfrey, Tony; Reichelt, Jon. *Industrial Enzymology: The Application of Enzymes in Industry.* (New York: The Nature Press, 1983).

Graham, David Y. "Ins and outs of pancreatic enzymes and adjuvant therapies." *Journal of Pediatric Gastroenterology and Nutrition* 3:S120–26 (1984).

Grossman, M.I.; Greengard, H.; Ivy, A.C. "The effect of dietary composition on pancreatic enzymes." *American Journal of Physiology* 139:676–82 (1942).

Grossman, M.I.; Greengard, H.; Ivy, A.C. "On the mechanism of the adaptation of pancreatic enzymes to dietary composition." *American Journal of Physiology* 141:38–41 (1944).

Guyton, Arthur C. M.D. *Textbook of Medical Physiology, 8th edition.* (Philadelphia: W.B. Saunders Company, 1991).

Heinbecker, Peter. "Further studies on the metabolism of Eskimos." *The Journal of Biological Chemistry* 93:327–36 (1931).

Howell, Edward, M.D. *Enzyme Nutrition: The Food Enzyme Concept.* (Wayne, NJ: Avery Publishing Group, Inc., 1985).

Howell, Edward, M.D. *Food Enzymes for Health and Longevity* (2nd Edition. Lotus Press. March 1994).

Ivy, A.C.; Schmidt, C.R.; Beazell, J.M. "On the effectiveness of malt amylase on the gastric digestion of starches." *The Journal of Nutrition* 12: 59–83 (1936).

Jackson, P.G.; Lessof, M.H.; Baker, R.W.R.; Ferrett, J.; MacDonald, D.M. "Intestinal permeability in patients with eczema and food allergy." *Lancet* I: 1285–6 (1986).

King, Leighton S. "Is the human pancreas hypertrophied?" (Unpublished report, 1980).

Klingerman, Alan E. "Relative efficiency of a commercial lactase product." *American Journal of Clinical Nutrition* 51:890–3 (1990).

Lehinger, Albert L. *Principles of Biochemistry.* (New York: Worth Publishers, Inc., 1982).

McLean, E.; Ash, R. "The time-course of appearance and net accumulation of horseradish peroxidase presented orally to Rainbow Trout, Salmon Gairdner." *Comparative Biochemistry and Physiology* 88A:507–10 (1987).

Memmler, Wood. *Structure and Function of the Human Body, Second Edition* (J.B. Lippincott Co., 1977).

Miller, Benjamin F., M.D.; Keane, Claire Brackman, RN. *Encyclopedia and Dictionary of Medicine, Nursing, and Allied Health, 4th Edition.* (W.B. Saunders Company).

Pottenger, Francis M. Jr., M.D. *Pottenger's Cats: A Study in Nutrition* (LaMesa, CA: Price-Pottenger Nutrition Foundation, 1983).

Rabinowitch, I.M.; Smith, F.C. "Metabolic studies of Eskimos in the Canadian Arctic." *The Journal of Nutrition* 12:337–56 (1936).

Reed, Patsy Bostick. *Nutrition and Applied Science (West Publishing Co., 1980).*

Schwimmer, Sigmund, Ph.D. *Source Book of Food Enzymology.* (Westport CN: The AVI Publishing Company, Inc., 1981).

Smyth, R.D.; Brennan, R.; Martin, G.J. "Systemic biochemical changes following oral administration of the proteolytic enzyme bromelain." *Archives of International Pharmacodynamics* 136:230–6 (1962).

Van De Graaff & Fox. *Concepts of Human Anatomy and Physiology Anatomy. William C. Brown Publisher.*

Walker, Norman W., D.Sc., Ph.D. *Colon Health The Key To a Vibrant Life.* (Norwalk Press, 1977).

Wilson, Fisher, and Garcia, *Principles of Nutrition, Fourth Edition.* (John Wiley & Sons, 1979).

References for Chapters Eleven and Twelve

1. Abderhalden. *Fermentforschung.* 15:93–120, 1936.
2. Alexander, *et al. J. Clin Invest.* 15:163–67, 1944.
3. Bartos and Groh. *Proc. Soc. Exp. Biol. Med.* 37:613–615.
4. Bienenstock, J. *Nutrition Reviews* 42:105–8, 1984.
5. Borgstrom, G. *Principles of Food Science.* Macmillan Co. 1968.
6. Brown, Pearce, & Van Allen. *J. Exp. Med.* 42:163–78, 1925.
7. Burge & Neill. *Amer. J. Physiol.* 63:545–47,1923.
8. Couey, Dick. *Nutrition for God's Temple,* Mellon Press, 1993.
9. Dixon, M. & Webb, E. *Enzymes.* Academic Press, 1979.
10. Dutta, S., Rubin, J., Harvey, *J. Gastroenterology* 84: 476–82, 1983.
11. Fiessinger et al. *Enzymologia* 1:145-50, 1936.
12. Fisher, *Proc. Soc. Exp. Biol. Med.* 29:400–494, 1932.
13. Fuller, DicQie. *Health Practitioner Manual.* 1991.
14. Gardner, M. *Annual Review of Nutrition* 8:329–50, 1983.
15. Graham, D. *J. Pediatric Gastroenterology and Nutrition* 3:120–6, 1984.
16. Grossman, M., Greengard, H., Ivy, A. *Amer. J. Physiology* 141:38–41, 1944.
17. Harrison, Denton, and Lawrence *Brit. Med. J.* 1:317–19, 1923.
18. Heinbecker, P. *J. Biological Chemistry* 93:327–36, 1931.
19. Howell, E. *The Status of Food Enzymes in Digestion and Metabolism.* Avery Publication, 1946.
20. Howell, Edward. *Food Enzymes for Health Longevity.* Lotus Press, 1994.
21. Howell, E. *Enzyme Nutrition.* Avery Publishing Co. 1985.
22. Ivy, A., Schmidt, C., Beazell, J. *J. of Nutrition* 12:59–83, 1936.
23. John. *J. Amer. Med. Assoc.* 101:184–7, 1926.
24. Kligerman, A. *Amer. J. of Clinical Nutrition* 51:890–3, 1990.
25. Kulvinskas, Viktoras. *Survival Into the 21st Century.* Omangod Press, Fairfield, Iowa, 1975
26. Mayer. *Bull. Johns Hopkins Hospital.* 64:246–7, 1929.
27. McCaughan. *Amer. J. Physiol.* 97:459–66. 1931.
28. Neilson & Terry. *American J. Physiol.* 15: 406–10, 1948.
29. Northrop. *J. Biol. Chem.* 32:123–6, 1917.
30. Reid, Quigley & Myers. *J. Biol. Chem.* 99:615–23, 1933.
31. Santillo, Smokey. *Food Enzymes.* Hohm Press, 1993.
32. Wasteneys. *Biochem. J.* 30:1172–82, 1936.
33. Wolf & Ransbager. *Enzyme Therapy.* Regent House, 1972.
34. Zucker *et. al. Amer. J. Physiol.* 102:209–21, 1932.